CW00695862

THE
KEW GARDENS
CHRISTMAS
BOOK

The Kew Gardens Christmas Book

Jenny Linford

Kew Publishing
Royal Botanic Gardens, Kew

Contents

RECIPES

INTRODUCTION

At the Royal Botanic Gardens, Kew, our purpose is to research and protect biodiversity. *The Kew Gardens Christmas Book* celebrates the important role that the natural world plays in this much-loved festival. In its pages, you'll find botanical facts, Christmas folklore, history, stories about Kew, plus some festive recipes, all interwoven with beautiful botanical illustrations from Kew's archive.

Each December sees Christmas celebrated widely around the world. While families in the UK traditionally mark the occasion with a splendid roast turkey, in the southern hemisphere, where it is summertime, Christmas is marked with parades, alfresco dining and barbecues. If we were to think of the Christmas festival itself as a plant, then it would be one which is long-lived, with deep roots and a great capacity to adapt to different environments. A look at the history of Christmas reveals echoes of ancient, pagan festivals marking the winter solstice on the shortest day of the year in December. There was a deep symbolic significance to the fact that the dark, cold days were ending and the cycle of the year would

[title page] Mistletoe (*Viscum album*) from *A Curious Herbal*, Elizabeth Blackwell, 1737.

[previous page] Nutcracker by Willem Wenckebach, 1898.

[opposite] Robin on a snowy branch by Theo van Hoytema, 1878–1910.

bring warmth, light and growth once again. Significant winter festivals included the Roman festival of Saturnalia and the Mithraic winter solstice festival Dies Natalis Solis Invicti, which marked the birthday of the unconquered sun. In the Bible and early Christian writings, no date is given for Christ's birth. The Christian story of the birth of Jesus overlaid a long history of winter solstice festivals, with the symbolism of light overcoming darkness as resonant to Christianity as it is to other religions.

A striking aspect of the much-loved festival of Christmas is the important role played in it by plants and animals. This is the time of year when we bring trees – real or artificial – into our homes and hang them with lights and special ornaments. In the UK, Christmas is when greenery comes indoors as we decorate our houses with sprigs of evergreen mistletoe, branches of prickly leafed, red-berried holly and trailing strands of ivy. In other parts of the world, different, local plants are associated with it. In the Falkland Islands, people enjoy the Christmas bush (*Baccharis magellanica*) which flowers in December bearing a mass of tiny white flowers. The fragrant white Christmas orchid (*Epidendrum ciliare*) of the British Virgin Islands, the yellow-flowering Christmas blossom (*Senna bacillaris*) of Montserrat and the Christmas palm (*Pseudophoenix sargentii*) on the Turks and Caicos Islands are other examples of plants linked to the festive season. From robins to reindeer, birds and animals have become part of how we portray Christmas in pictures, cards, stories, poems, films and TV shows, while (as part of our fantasy Christmas) a picturesque blanket of white snow sets the scene.

Sparrows on a snowy berry bush by Ohara Koson, 1900–45.

The festive season is very much a time of feasting. Over the centuries, certain culinary animals and birds and plant-based foods, from spices such as cinnamon or ginger to raisins and clementines, have come to be particularly associated with it. The traditional recipes these ingredients are used in – such as Christmas pudding, gingerbread or Christmas fruit cake – are a pleasurable, evocative aspect of this special time of year.

The nineteenth century saw Christmas become increasingly focused on the domestic home, and indeed, we now think of the period as a time in which we gather together with our families. For Kew's botanists, the practical reality of exploring the world on long voyages to seek out plants meant that Christmases were often spent far away from home, sometimes out at sea or in remote, inhospitable environments. One particularly striking account comes from the Victorian botanist and one-time Kew director Sir Joseph Dalton Hooker. His first major botanical expedition was on the HMS *Erebus* alongside HMS *Terror* as part of Captain James Clark Ross's Antarctica expedition (1839–1843). The Christmas of 1841 saw the expedition stuck in the unchartered expanse of the frozen Southern Ocean. Undaunted, however, Hooker and the crew threw themselves into the festive spirit, carving out settlements in the ice which they called Erebus Town and Terrorville, in which they could celebrate. In a letter he wrote to his sister, Hooker gives a vivid sense of the scale of what they'd constructed:

> *The snow was soon cut up into lanes & walks; after which, Saloons were trenched out, with seats all round, for dancing, – racing rounds & apartments for refreshments, – all ready for Christmas Eve. . . . I next devised some production on a grand scale, nothing short of a Sphynx, to be hewn out, just off the main road between Erebus Town & Terrorville . . . this Sphynx was to be 7 feet high & we labored many hours with great diligence . . . when a gust of wind came, bump went the Erebus against the floe:– there was a shock like an earthquake, – tottered the Sphynx on its very base & then gradually heeling over, crash came it down . . . & we were left kicking and sprawling in the ruins.*

Despite the mishap with the Sphynx, a merry time celebrating Christmas was had by Hooker and his companions.

Christmas is a festival we enjoy observing at Kew Gardens. The Christmas tree, which today is such a central part of people's Christmas celebrations, has its links to the Gardens. Queen Charlotte, the wife of King George III, was a keen botanist who often resided at the royal palace at Kew Gardens. She played an early, influential part in popularising the Christmas tree – a German custom – in the UK (see page 16). It is apt, therefore, that at Wakehurst, Kew's wild botanic garden in Sussex, our festive celebrations include decorating the UK's tallest living Christmas tree, a magnificent redwood. Visiting Kew Gardens to see the popular annual Christmas at Kew light trail – which illuminates the Gardens in a beautiful, imaginative way – has become a beloved Christmas tradition for many families.

Taking place as it does in December, when the calendar year draws to an end, Christmas has long been considered a time of reflection as well as one of celebration. In the Christian tradition, it is a period in which to think of others – to be kind, generous and giving. This is the message at the heart of Charles Dickens's influential story *A Christmas Carol* (1843), in which Scrooge the miser transforms into Scrooge the philanthropist. The natural world and the human world are closely interdependent – and it is well worth pausing to appreciate this special relationship.

Christmas cactus (*Schlumbergera truncata*) from *La Belgique Horticole*, Édouard Morren, 1866.

Pinus sylvestris

Evergreens

The bringing of greenery into the home for midwinter festivals is a practice thought to date back to long before Christianity. In pre-Christian rituals, the winter solstice was celebrated as the time when the sun began a new cycle of life, seeing off the darkness of winter. The Romans brought branches of bay (*Laurus nobilis*) into their homes as part of the festival of Saturnalia in mid-December. In the UK, the plants traditionally associated with Christmas are evergreen ones, among them holly (see page 25), mistletoe (see page 28) and the coniferous Christmas tree (see page 16). In folk customs, evergreen plants have come to symbolise the life force, immortality and fertility. We use the term 'evergreen' to refer to something that renews itself or remains constant. In botany, the term 'evergreen' is used to describe plants with foliage that remains functional for more than one growing season. Evergreen plants occur in a variety of climates ranging from the warm tropics through to the taiga or boreal forests found in the northern hemisphere between the temperate forests and the tundra, in the cold subarctic region just south of the Arctic Circle.

Deciduous plants shut down during the cold, dark winter months. They withdraw chlorophyll – the green pigment which is key to photosynthesis – from their leaves and shed their leaves all at once. Evergreens also shed their leaves, but they do this gradually over an extended period of time. Among evergreens, conifers are characterised by their small, slender, tough and leathery leaves, which we call needles – very different from the broader, softer leaves found on deciduous trees. These distinctive leaves are specially adapted to help conifers cope with cold winters. The needles are covered with a waxy coating that prevents water

Scots pine (*Pinus sylvestris*) from *Medical Botany*, John Stephenson, 1835.

loss through transpiration. The distinctive pointed shape of many coniferous trees and the shape of their branches allow snow to slide off. Northern conifers are able to alter their internal biochemistry to avoid freezing and can cope with high levels of dehydration. Their thick bark forms a protective layer, and their pine cones protect the seeds inside them. By retaining their foliage throughout winter, conifers are already fully leafed when spring begins. During the dark, cold winter months, the green foliage offered by conifers and other evergreens has long been regarded by human beings as a hopeful sight.

THE CHRISTMAS TREE

Of the various plants associated with Christmas, the most striking and central is, of course, the Christmas tree. Bringing a tree – usually an evergreen conifer – into the house and decorating it has become a Christmas ritual enjoyed in many homes. And it is not just in homes that we find Christmas trees. Institutions and businesses similarly enter into the Christmas spirit. At Wakehurst in 2022, 1,800 illuminating low-energy, LED light bulbs were used to decorate the UK's tallest living Christmas tree, a magnificent redwood (*Sequoiadendron giganteum*) measuring 37 metres (121 ft). The White House, the residence of the American president, has an official White House Christmas tree. In 1961, First Lady Jacqueline Kennedy began the practice of choosing a theme for the White House Christmas tree, looking to Tchaikovsky's ballet *The Nutcracker* for inspiration.

There is also a strong tradition of public Christmas trees in iconic locations. In New York, there's the Rockerfeller Center Christmas Tree, the lighting up of which is broadcast live. In London, the Trafalgar Square Christmas Tree is set up in this central square and adorned with lights. The tall Norway spruce is a gift each year from the people of Norway to the people of the UK

Christmas tree (*Abies nordmanniana*) from *Neerland's Plantentuin*, Cornelis Antoon Jan Abraham Oudemans, 1865.

MEIJER'S PRENTEN
2e Serie.

No. 28

HET LOT VAN EEN KERSTBOOM.

Een jonge, groene denneboom
Stond op den heuvel bij den stroom.

De wind huilt door het donker woud.
En breekt het broze en dorre hout.

Tot op zek'ren kouden dag
De spar een man verschijnen zag,

Die zonder omslag, kort en goed,
Hem omhakt en dus vallen doet.

Hij legt hem op een groote sleê
En voert aldus ter markt hem meê.

Dáár ziet hem een aanzienlijk heer,
En zegt „dat 's juist wat ik begeer."

Hij maakt er nu een kerstboom van,
En siert hem op zooveel hij kan.

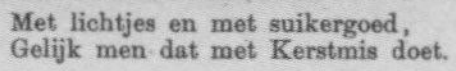

Met lichtjes en met suikergoed,
Gelijk men dat met Kerstmis doet.

Nu 't suikergoed is opgepruimd
Wordt ook de boom weer weggeruimd,

Tot spijt van heel der kind'ren schaar
Want Kerstmis komt maar eens in 't jaar.

Daar ligt nu 't boompje in den hof
Bij asch en vuilnis, zand en stof,

Een arme vindt hem op het land,
En neemt hem meê voor kachelbrand.

– a token of gratitude for support shown during the Second World War. The custom started in 1942 when, during a raid on the island of Hisøy, a Norwegian resistance fighter called Mons Urangsvåg decided to cut down a Norway spruce and send it as a gift to his king, Haakon VII, who had escaped to the UK following the Nazi occupation of his country and set up a government-in-exile in London. The king's Norway spruce was then put on display in Trafalgar Square, though without lights, owing to blackout rules. The war ended, and since 1947, a Norway spruce has been sent to London from Norway each year.

While there is a long history of trees being decorated for Christmas, there are a number of theories as to when and where this tradition began. There are early reports of decorated trees in Freiburg in 1419, Tallinn in 1441 and Riga in 1510, each variously claimed to be the 'first' Christmas tree. One theory links the origin of the Christmas tree to the medieval mystery plays, which featured a Paradise Tree decorated with apples and communion hosts, and representing the Tree of the Knowledge of Good and Evil in the Garden of Eden. Another theory suggests that the practice of dressing trees originates in an ancient Central Asian winter festival.

Various references show that the custom of having a Christmas tree became increasingly established in Europe, with Germany playing a central role in the story. There is a record of a Christmas tree in Strasbourg Cathedral in 1539. In 1554, rather tellingly, the felling of trees for Christmas was banned in Freiburg. By the sixteenth century, it had become a custom for German trade guilds to have decorated trees in their guildhalls at Christmas. In a popular German legend, the Christmas tree is associated with the church reformer Martin Luther (1483–1546). While walking through the woods on a dark winter's night, he is said to have been inspired to bring home a tree to his family, setting candles on it to evoke the bright stars he had seen shining in the heavens above. The Christmas tree was seen as a Protestant alternative to the

The fate of a Chrtismas tree. **Print by De Ruyter and Meijer, Amsterdam, 1874.**

Catholic manger scene. An account shows that by 1605, inhabitants of Strasbourg were setting up 'fir trees' in their parlours which they then adorned with 'roses cut out of many-coloured paper, apples, wafers, gold-foil, sweets, etc'. In Germany during the seventeenth and eighteenth centuries, it became fashionable in aristocratic and wealthy circles to have a Christmas tree, or if not a whole tree, large branches cut from a yew tree (*Taxus baccata*). The Romantic poet Samuel Taylor Coleridge wrote a popular, much-published account of a yew branch ceremony he had witnessed on a visit in 1798 to Mecklenburg-Strelitz in Germany at Christmastime.

Traditionally, there was a sense of anticipation and excitement attached to the 'unveiling' of the Christmas tree in Germany. It was only on Christmas Eve that the candles on the tree were first lit and the children – who until then had not been permitted in the room – were allowed to see the tree in all its glory. The acclaimed German poet Goethe described this moment in his novel *The Sorrows of Young Werther* (1774), writing evocatively of 'the unexpected opening of the door and the appearance of the marvellous tree with its wax candles, sweets and apples'. For centuries, the decorations placed on Christmas trees alongside lighted candles included edible ones: apples, gilded nuts, sweets and gingerbread. Ornamental Christmas tree decorations were also made from an edible, plant-based gum called tragacanth. The gum is made from shrubs of the *Astragalus* genus (mainly *A. gummifer* and *A. microcephalus*), and has the useful quality of being mouldable into shapes such as figurines.

In the UK it is the royal family who are seen as playing an important part in introducing and spreading the German tradition of the Christmas tree. King George III's wife Queen Charlotte – for whom Kew Palace was a summer home and whose cottage can be seen in the grounds at Kew Gardens – is credited with introducing the first Christmas tree to the UK. She had grown up in the duchy of Mecklenburg-Strelitz and introduced the yew branch ceremony

Hand-coloured print of a winter scene, c. 1860.

E. B. & E. C. KELLOGG, 245 MAIN ST. HARTFORD, CONN. F. P. WHITING, 87 FULTON ST. NEW-YORK

THE CHRISTMAS TREE.

169.

to her young family and the royal court. At Windsor Castle in 1800, she held a large Christmas party for the children of leading families in Windsor. As a centrepiece, rather than simply having a yew branch, she decided to have a whole yew tree and decorate it. Her biographer John Watkins, who was present at the party, described the scene as follows:

> *In the middle of the room stood an immense tub with a yew tree placed in it, from the branches of which hung bunches of sweetmeats, almonds, and raisins in papers, fruits and toys, most tastefully arranged, and the whole illuminated by small wax candles. After the company had walked around and admired the tree, each child obtained a portion of the sweets which it bore together with a toy and then all returned home, quite delighted.*

Many of the Christmas customs we continue to enjoy today came to the fore in the Victorian era. Queen Victoria and her German-born husband, Prince Albert, are credited with doing much to popularise the Christmas tree. Prince Albert introduced his young family to its joys, and in 1847, thinking back to his childhood with his brother, he wrote: 'I must now seek in the children an echo of what Ernest and I were in the old time, of what we felt and thought; and their delight in the Christmas-trees is not less than ours used to be.' In December 1848, the widely read *Illustrated London News* featured an influential illustration of Queen Victoria and Prince Albert with their family at Windsor Castle, gathered around a tabletop Christmas tree, beautifully decorated with ornaments and candles. Queen Victoria was hugely popular, and this example set by the royal family of having a Christmas tree was followed by the British public with enthusiasm. Interestingly, this festive image of Queen Victoria and her family also played a part in promoting the Christmas tree in North America: it was reproduced there in 1850 in *Godey's Lady's Book*, a widely circulated magazine. The second half of the nineteenth century saw the Christmas tree become

Christmas Tree at Windsor Castle by J. L. Williams, *The Illustrated London News*, 1848.

increasingly accepted as part of Christmas in the US.

In the northern hemisphere, Christmas trees are usually evergreen conifers. Conifers are an ancient group of plants, with fossil records dating back more than 300 million years. They are part of a group of plants known as gymnosperms, which means 'naked seeds', as opposed to angiosperms, in which the seeds are enclosed. Virtually all conifers produce cones which bear the seeds.

The term 'fir trees' is often used for conifers but it's misleading, as the genus *Abies* (the true fir) contains only 47 of the 615 species of conifers worldwide. Among the conifer species popularly used as Christmas trees are the Norway spruce (*Picea abies*), blue spruce (*Picea pungens*), white spruce (*Picea glauca*), Nordmann fir (*Abies nordmanniana*), Fraser fir (*Abies fraseri*), white or Weymouth pine (*Pinus strobus*), Scots pine (*Pinus sylvestris*) and lodgepole pine (*Pinus contorta*). The Norway spruce and Fraser fir are noted for their scent, while the Nordmann fir holds onto its needles well. The Nordmann fir is indigenous to the mountains south and east of the Black Sea, named after the Finnish botanist Alexander von Nordmann. It is a very popular Christmas tree, valued for its attractive foliage and retentive needles. Nowadays, most Christmas trees are specially grown on Christmas tree farms. It takes around eight years for a Christmas tree to reach market size. In the wild, the species we use as Christmas trees can grow to great heights, with the Nordmann fir and Norway spruce reaching 60 metres (197 ft). Of these Christmas conifers, the Scots pine is the only pine tree native to the UK. It can grow to 35 metres (115 ft) and live for up to 700 years. Many of them are found growing in the Scottish Highlands, where they support rare species including the red squirrel and the pine marten.

While the economic demand for Christmas trees ensures that they are widely cultivated, not all conifers are thriving. The rarest conifer tree in the world is the Baishan fir (*Abies beshanzuensis*), which is considered Critically Endangered by the International

Union for Conservation of Nature. Just a few wild trees of the species remain today, all growing within a single site on Mount Baishanzu in China. Kew scientists are studying the relationship between the tree and a rare species of fungus found growing on its needles. As part of Kew's work to help retain biodiversity, young saplings of the Baishan fir are being grown in the Kew nurseries and also in new areas in China, planted by Kew's partners at the Chinese Academy of Forestry.

HOLLY

With its glossy green leaves and bright scarlet berries, common holly (*Ilex aquifolium*) is one the plants we traditionally use to decorate our homes for Christmas. Indeed, the first line of a popular Christmas carol exhorts us to 'Deck the halls with boughs of holly'.

Holly was considered a special plant long before Christianity. Its vitality in the cold, dark winter months, retaining its leaves while other trees lose theirs, meant it was valued by the Druids as a magical plant, offering the hope of new life at the winter solstice. It is a plant with much folklore attached to it. In Norse mythology, holly is associated with Thor (god of thunder) and Freya (goddess of love and fertility). It was considered to have special protective and healing powers and was planted near houses to protect them from lightning. While cutting branches of holly to decorate homes was acceptable, the felling of holly trees was long considered to be unlucky. In the medieval poem *Sir Gawain and the Green Knight*, the mysterious Green Knight who comes to a winter feast at King Arthur's court carries a branch of holly 'that shows greenest when all the groves are leafless'. The figure of the Green Knight is interpreted by some as a representation of nature and rebirth. In neopaganism, one finds the legend of the Oak King (representing summer) and the Holly King (representing winter), who battle each

ILEX Aquifolium. HOUX Commun. *pag. 1*

Redouté pinx.t

other for supremacy each year.

In the Christian tradition, holly is associated with Christ. The sharp thorns on its leaves represent the crown of thorns he wears on the cross and its red berries his blood. In the Christmas carol *The Holly and the Ivy*, a connection is made between Jesus and the holly tree in this verse:

The holly bears a prickle,
As sharp as any thorn,
And Mary bore sweet Jesus Christ
On Christmas Day in the morn.

To this day, holly continues to be seen as an element of Christmas, from the holly wreaths on front doors to the sprig of holly placed on top of the Christmas pudding.

Holly has a special place at Kew Gardens. Visitors to the Gardens can stroll along Holly Walk, near the Temperate House. Planted in 1874 by Sir Joseph Hooker, this is one of Europe's most comprehensive holly collections, containing more than 30 different species.

Behind the scenes, Kew is also working to secure the genetic diversity of holly as part of the UK National Tree Seed Project (UKNTSP), for which seeds from native trees and shrubs are being collected and stored. At the Millennium Seed Bank, Wakehurst, there are now 82 holly collections (comprising of 69 UK collections and 13 collections from MSB partner countries), totalling 412,553 seeds. It's important to know if the seed is alive and how to germinate it, so germination tests are carried out on the stored seed. Many seeds have a dormancy mechanism, which means that the seed will only germinate when the conditions favour the seedling's survival. For many tree species in the UK, the breaking of dormancy is linked to temperature, with seeds requiring a period of cold before they germinate, so the natural, seasonal cycle of a cold winter followed by a warmer spring is perfect. Getting the

Holly (*Ilex aquifolium*) from *Traité des arbres et arbustes que l'on cultive en France en pleine terre*, Henri Louis Duhamel du Monceau, 1801.

collected holly seeds to germinate proved surprisingly complex and time-consuming for such a common species; however, after three and a half years of patient testing and waiting, the holly seeds did germinate successfully, meaning we now understand more about this plant which plays an important part in our festive celebrations.

MISTLETOE

One of the plants now associated with Christmas is mistletoe (*Viscum album*), with its distinctive pale-green leaves and white berries (known as drupes). Mistletoe is an evergreen hemiparasitic plant that grows on trees, often apple trees in the UK, but also on poplars and limes. It coexists with its host tree without killing it, though it may weaken it. Mistletoe is spread by birds such as the mistle thrush, whose name reflects its fondness for the plant's berries. The seeds inside the berries are coated in a tacky substance called viscin, which allows the regurgitated or excreted seeds to stick and attach to a potential host tree. From a botanical point of view, mistletoe is an intriguing plant as it contains 25 times more DNA in each cell than we have in ours. As part of the collaborative Darwin Tree of Life project, for which Kew has partnered with other organisations including the Wellcome Sanger Institute and the University of Edinburgh, mistletoe's large genome has been sequenced in full.

Traditionally, a bunch of mistletoe is hung in the home during the festive season, and anyone who stands underneath it may be kissed. As with many traditions, the exact origins of this custom are not known, but by Victorian times the practice of decorating houses with mistletoe and kissing under it had become well established. The American author Washington Irving, credited with popularising Christmas customs in the US through his writing, described the practice in England in his bestseller *The Sketch Book* (1819–1820): 'The mistletoe is still hung up in farm-houses and

Mistletoe (*Viscum album*) from *Köhler's Medizinal Pflanzen*, Franz Eugen Köhler, 1887.

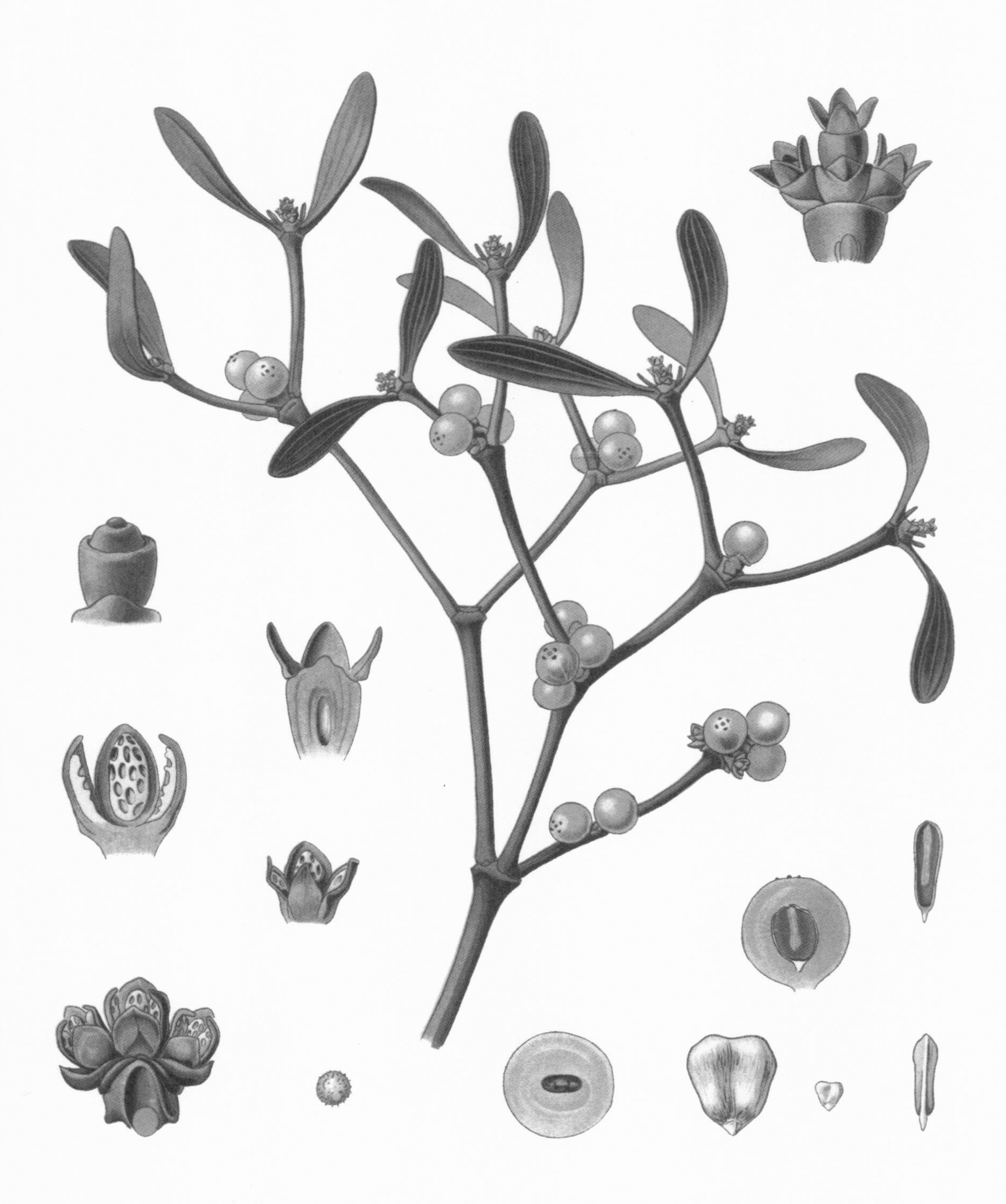

Valentine, print by anonymous artist, 1880.

kitchens at Christmas, and the young men have the privilege of kissing the girls under it, plucking each time a berry from the bush. When the berries are all plucked the privilege ceases.' The English novelist Charles Dickens, who similarly did so much to popularise a vision of domestic Christmas in the UK, wrote about a jovial festive scene of kissing under the mistletoe in *The Pickwick Papers* (1836): 'Mr Pickwick, with a gallantry that would have done honour to a descendant of Lady Tollimglower herself, took the old lady by the hand, led her beneath the mystic branch and saluted her in all courtesy and decorum.' In the 1950s we find the young singer Jimmy Boyd crooning about how he caught mommy kissing Santa Claus 'underneath the mistletoe last night'.

Mistletoe has a long history in mythology and folklore and is rich in plant magic associations. It is perhaps not surprising that it was attributed with special powers – growing as it does on trees (rather than being rooted in the earth as other plants are), and producing fruit in winter months. In Norse mythology, mistletoe is the one plant in the world with the power to kill Baldur, the beloved son of Odin and the goddess Frigg, and a dart of mistletoe is used by the scheming Loki to cause Baldur's death. As anyone who grew up reading René Goscinny and Albert Uderzo's *Asterix* books will know, the Druids are said to have attributed mistletoe with magical powers. Pliny the Elder, in his *Natural History* (CE 77), wrote that the Druids 'held nothing more sacred than the mistletoe and the tree that bears it'. He went on to say that mistletoe rarely grew on oak, but when it was found there, a special ceremony would be held in which a Druid clad in white robes would ascend the oak tree to cut the mistletoe with a golden sickle. He wrote, 'It is the belief with them that the mistletoe, taken in drink, will impart fecundity to all animals that are barren, and that it is an antidote for all poisons.' In reality, though, mistletoe berries are toxic and should not be eaten. This long association with fertility may underlie the practice of kissing under the mistletoe, which still takes place to this very day.

IVY

The climbing plant known as ivy (*Hedera helix*) has discreetly twined its way into our Christmas celebrations. Its distinctively shaped, pointed leaves are twisted into wreaths alongside branches of prickly holly and used to decorate our homes, adorning mantelpieces and featuring in floral centrepieces.

Ivy – which, significantly, is an evergreen – has long been associated with winter festivals. Growing in the temperate regions of the world including Europe, ivy was valued for the fact that it stays green and leafy through the cold, dark winter months when

Ivy (*Hedera helix*) from *Types de Chaque Famille et des Principaux Genres des Plantes Croissant Spontanément en France*, François Plée and Jean-Baptiste Bailliere, 1864.

so many other plants die back. Indeed, the ivy's flowers are an important source of winter nectar for bees and its berries a useful food for birds. In ancient Roman times, Saturnalia, which took place from 17 to 23 December, was a celebration of Saturn, the god of agriculture, seed and sowing. During this festival, evergreens were used to decorate temples and homes. Another Roman god also associated with Saturnalia – a period of exuberant celebrations when much drinking took place – was the god of wine, known to Romans as Bacchus and to the Greeks as Dionysus. Bacchus and his followers, the maenads, were often depicted wearing wreaths made from ivy. Among the various powers attributed to ivy was the useful one of preventing drunkenness: all you had to do was drink your wine from a goblet made from ivy wood. Such drinking vessels made of ivy were also thought to protect against poisoning. In a further association with alcohol and merrymaking, poles covered with ivy were hung outside medieval alehouses as a sign to travellers that there was a hostelry there.

Presumably because of ivy's tenacity in clinging on, via its adhesive aerial roots, to the surface on which it grows up (such as tree trunks and walls), the plant came to symbolise fidelity and friendship. Its association with Christmas is apparent in the popular carol *The Holly and the Ivy*. Ivy has long been used to decorate homes and churches for Christmas, with its dark berries sometimes painted red. The English poet John Clare (1793–1864) wrote of ivy being picked and its berries coloured 'with whiting'. In his book *Flora Britannica* (1996), Richard Mabey gives an account of a Shropshire farmer who during the 1930s gave each cow a sprig of ivy before noon on Christmas Day in order to help fend off the Devil during the coming year. In another superstition, an ivy leaf set to float on water on New Year's Eve and left untouched until Twelfth Night was used to predict the fate of the person who picked it. If the leaf stayed fresh, the prediction was for good times, but if it withered or blackened, troubles were expected.

The Yule log

Among the historic Christmas customs centred on trees is the ceremonial gathering and burning of the Yule log. The word 'Yule' comes via Old English from the Old Norse word *jól*, meaning a winter celebration. It originally referred to a pre-Christian winter festival held by the Germanic peoples and possibly connected to the myth of Odin (or Wotan) and the Wild Hunt. Over time, however, the word Yule has come to mean the Christmas period.

The Yule log custom involved the bringing of a special, large log, known as the Yule log, into the home on Christmas Eve. There it was placed on the fireplace and lit with a piece of the previous year's Yule log that had been carefully saved. In some versions of the tradition, the Yule log was kept burning through the Twelve Days of Christmas until Twelfth Night on 6 January. It was important not to burn up the log completely, as a piece needed to be reserved for the following Christmas. It was a ceremony practised in numerous European countries, and often the Yule log was blessed or sprinkled with wine and breadcrumbs, symbolising Communion. The Yule log was often attributed with special powers to bring good fortune and protect against evil spirits. The symbolism of creating light and warmth during the darkest time of the year through the lighting of a fire is a powerful one.

The English poet Robert Herrick (1591–1674) described the custom in two stanzas of his poem *Ceremonies for Christmas*:

Come, bring with a noise,
My merry, merry boys,
The Christmas Log to the firing;
While my good Dame, she
Bids ye all be free;
And drink to your heart's desiring.

[above]
***The Yule Log* by Albertine Randall Wheelan, c. 1908.**

[below]
***Yule Log Lugged*, greeting card c. 1905.**

Happy Christmastide
Come bring with a noise My merrie merrie boys The Christmas log to ye firing
Albertine Randall Wheelan

Christmas Greetings.
Old times, Old friendships.

With the last year's brand
Light the new block, and
For good success in his spending,
On your Psaltries play,
That sweet luck may
Come while the log is a-tinding

In an age when many homes no longer have working fireplaces, the practice has declined. The idea of the Yule log, however, lives on in a smaller, edible form which can easily be enjoyed – the French *bûche de Noël* (Christmas log), a sponge cake shaped and decorated to resemble a picturesque wooden log.

The Glastonbury Thorn

A tree with a special resonance during the Christmas season is the Glastonbury Thorn, found in Somerset. According to Arthurian folklore, Joseph of Arimathea, a wealthy follower of Jesus Christ who buried his body after the Crucifixion, journeyed to the UK carrying the Holy Grail in order to bury it in a secret spot just below Glastonbury Tor. While at Glastonbury, he climbed up Wearyall Hill, stuck his wooden staff into the ground and rested. When he awoke, his staff had miraculously taken root, flowered and turned into the hawthorn tree (*Crataegus monogyna* 'Biflora') known as the Glastonbury Thorn or Holy Thorn. In some versions of the story, the staff was made with the wood of the Cross. Joseph and his followers are said to have established a church at Glastonbury, laying the foundations for what became the abbey there.

Unlike ordinary *Crataegus monogyna*, the Glastonbury Thorn flowers twice a year, once at Easter and once at Christmas, both

Glastonbury Thorn (*Crataegus monogyna*) from *Icones Plantarum Sponte Nascentium in Regnis Daniæ et Norvegiæ*, Georg Christian Oeder, 1761.

special times in the Christian calendar, and was attributed with special powers. Botanically speaking, the capacity of this hawthorn variety to flower twice is indicated by its cultivar name, 'Biflora'. While the abbey was dissolved in 1539, the original Glastonbury Thorn continued to grow and flower at Christmas until it was cut down by one of Oliver Cromwell's soldiers during the English Civil War. However, cuttings from the Glastonbury Thorn were taken and propagated, so there are descendants from the original tree still to be found at Glastonbury.

A tradition of sending a sprig from the Holy Thorn to the reigning monarch at Christmas is said to have been established in the early seventeenth century, when James Montague, Bishop of Bath and Wells, sent a branch to Queen Anne, wife of King James I. The practice was revived in the early years of the twentieth century and continues to this day with the Holy Thorn cutting ceremony, when a flowering sprig is cut in mid-December in the grounds of St John's Church, Glastonbury (home to one of the Glastonbury Thorn's descendants), and then sent to the sovereign.

In 2010 there was outrage and dismay when vandals brutally hacked down the Glastonbury Thorn on Wearyall Hill. The tree had been the object of pilgrimage, with ribbons, prayers and offerings tied to its branches. Cuttings were collected from the severed branches by Kew horticulturists and taken to Kew Gardens where they were grafted onto hawthorn rootstock and nurtured. Three years later, one of the trees grown from this grafting at Kew was ceremonially planted at Wearyall Hill. Sadly, however, in 2019 this tree was chopped down and removed by the landowner. However, there are other known descendants of the Glastonbury Thorn to be found, including one in the New York Botanical Garden and a number in Kew Gardens.

Ornamental Christmas plants

In countries with cold, harsh winters, there has long been a sense of wonder attached to the idea of plants that flower at this time of year. In Christian folklore, we find stories of trees and plants that miraculously flower at Christmas (see Glastonbury Thorn, page 36). Inspired by the prophecy of Isaiah, a fifteenth-century German poem evoked this image in its first stanza:

A Spotless Rose is growing,
Sprung from a tender root,
Of ancient seers' foreshowing,
Of Jesse promised fruit:
Its fairest bud unfolds to light
Amid the cold, cold winter
And in the dark midnight

In Franconia (in northern Bavaria), in a custom dating back to the thirteenth century, cuttings from lilac (*Syringa vulgaris*), cherry (*Prunus avium*) and apple trees (*Malus domestica*) were taken on 4 December (the feast day of St Barbara) and brought into the warmth of the house so that they would bud and flower for Christmas.

In northern Europe, a number of plants are valued because they flower in December. The 'Christmas rose' is, in fact, *Helleborus niger* – a member of the buttercup family with white or pink-flushed flowers. Two species of cacti, *Schlumbergera truncata* (see page 10) and *Schlumbergera buckleyi*, are known as 'Christmas cactus' because they flower from late November to January. Pots of flowering bulbs, such as hyacinth (*Hyacinthus orientalis*) with its

Poinsettia (*Euphorbia pulcherrima*) from *Curtis's Botanical Magazine*, 1836.

heady scent, have become a popular Christmas decoration or gift. Another decorative plant that also grows from bulbs is amaryllis (*Hippeastrum*), which has a tall stem and large, striking blooms.

Originally from Mexico, the poinsettia (*Euphorbia pulcherrima*) – which the Aztecs called *cuitlaxochitl* – has become associated with Christmas in the northern hemisphere in the form of an indoor house plant. With its green leaves and scarlet bracts (modified leaves) which surround the flowers, it is valued for its festive colouring. It is also appreciated because, unlike fragile petals, bracts can last for months. A Mexican legend associated with the plant tells of a poor girl named Pepita, who went to church on Christmas Eve and picked a humble bunch of weeds to offer to the baby Jesus. When she placed the weeds on the altar, they miraculously flowered, and in Mexico, poinsettia is called *la flor de Nochebuena* – or the Christmas Eve flower.

Christmas wreaths

A circular wreath of greenery – often made from intertwined holly or pine branches and ivy – is a familiar sight on front doors during the Christmas period. There is a symbolic power to the shape of a circular wreath. The circle – a shape without a beginning or an end – is a universal symbol of many things, among them infinity and wholeness. The wreath itself has a venerable history of being associated with celebrations. Laurel wreaths were awarded by the ancient Greeks to winning athletes and to triumphant commanders by the Romans. Several European countries have a tradition of harvest wreaths. Made from harvested grains, fruits and nuts, the harvest wreath celebrated the successful gathering in of the crop and brought good fortune. In Scandinavia, to celebrate St Lucia's

Gathering evergreens and making wreaths, *Harper's Weekly*, 1858.

Day on 13 December, it's traditional for girls to dress as St Lucia – in a white gown, and with an evergreen wreath adorned with candles on their heads.

The Advent wreath is said to have been invented in 1833 by a Lutheran clergyman called Johann Wichern who ran a missionary school in Hamburg. In the Christian calendar, Advent is the period of preparation for the birth of Jesus Christ. In order to give the children he was teaching a visual representation of the Advent season, Wichern would place four white candles in a wreath and light one candle each Sunday. Advent wreaths, either home-made or shop-bought, are still a popular Christmas tradition in Germany.

The Christmas wreath draws on the long tradition of decorating homes with greenery for Christmas. The wreaths are decorated in a variety of ways – adorned with items such as berries, dried citrus fruit, bundles of cinnamon sticks, fir cones or baubles. Nowadays, commercially made wreaths are widely sold, but many families enjoy the process of foraging for greenery or picking it from their gardens to make their own wreath.

Wassailing

The custom of wassailing fruit orchards around Christmas and the New Year is a tradition with historic roots. The word 'wassail' comes from the Anglo-Saxon *waes hael*, meaning 'be in good health'. Through the centuries, the practice of drinking toasts from wassail bowls or cups became established. In villages and towns during the Twelve Days of Christmas, wassail cups were carried from house to house by wassailers offering toasts in exchange for food, drink or money.

In rural communities, wassailing became the offering of a toast by agricultural workers to their local fruit trees to encourage them to bear fruit plentifully. The seventeenth-century poet Robert Herrick, who chronicled rural traditions in his poetry, wrote about wassailing on a number of occasions:

> *Wassail the trees, that they may bear*
> *You many a plum, and many a pear:*
> *For more or less fruits they will bring,*
> *As you do give them wassailing.*

Wassailing became closely connected to cider making, taking place in the traditional cider counties of Devon, Gloucestershire, Herefordshire and Somerset. Villagers would make their way to the largest or most productive tree in their local apple orchard and

Scene 14
WEBB'S SCENES in HARLEQUIN JACK and the BEANSTALK
No 13
WINE
WASSAIL
London. Pub. by W. WEBB. 146. Old St. St Lukes.

A scene from Webb's toy theatre set for *Jack and the Beanstalk*, W. Webb, late nineteenth century.

pour cider or beer over its roots. Pieces of soaked toast or cake were placed in the branches of the tree for the robins, regarded as the trees' guardian spirits. The wassailers filled their cups with cider and tossed it into the branches, before refilling them to drink the cider and sing a toast to the tree. Villages in these counties are said to have each had their own version of a wassailing song. One historic song is as follows:

Here's to thee, old apple tree
Whence thou may'st bud and whence thou may'st blow
And whence thou may'st bear apples enow.
Hats full, Cups full, Bushel, Bushel Sacks full,
And my pockets full too!
Huzza!

Making loud noises to scare away evil spirits and ensure a good harvest was another part of wassailing, hence the firing of shotguns into the air, the blowing of cow horns and the beating of buckets.

A recent resurgence of interest in producing craft cider has seen the revival of wassailing taking part in the UK's apple orchards once again.

The robin

The European robin is a popular garden bird in the UK, much appreciated for its melodic song and its friendliness towards humans. Over time, it has become a bird which is particularly associated with Christmas. In Christian folklore there are a number of stories about this small, bold bird. In one legend, the robin seeks to comfort Christ on the Cross by removing one of the thorns from his crown of thorns. As the bird performs this charitable act, a

Christmas scene, 1874.

drop of Christ's blood stains his breast red. In another story, the bird helps to look after the baby Jesus in the stable in Bethlehem. While Joseph is out getting fuel for the fading fire, Mary requests help from the animals in keeping it alight so as to keep the infant warm. A little brown bird valiantly fans the flames with its wings, scorching itself and so gaining its distinctive red breast. In Welsh folklore, the kindly robin flew down to hell to bring water to the sinners and scorched its breast on the flames. The English poet William Wordsworth, in a poem of 1802, calls the robin 'the pious bird with the scarlet breast'.

The association of the robin with Christmas specifically is linked to the invention of the Christmas card by Sir Henry Cole, the founding director of the Victoria and Albert Museum. Significantly, Cole had also played a leading part in setting up the Uniform Penny Post system. Introduced in 1840, the Penny Post was a great stimulus to written communication, making it affordable and accessible. In 1843, Cole hit upon the simple but ingenious idea of sending a decorative, printed message of Christmas greetings

A robin on a branch of holly. Colour lithograph by Leighton Brothers after H. Weir, 1858.

by post to his many acquaintances. He commissioned a sketch of three generations of the Cole family raising a toast from the artist John Callcott Horsley. Cole then had the sketch printed on cards, which he posted to his friends after adding personalised messages of seasonal greetings. Throughout the Victorian era, the Christmas card grew in popularity. Many of the classic images we now associate with the festive season appeared on these early cards: snowy landscapes, churches, Father Christmas, evergreens, Christmas feasting and the robin. One popular theory as to why the robin appeared on Christmas cards is linked to the fact that the postman's uniform from 1793 had been scarlet in colour. Postmen in these bright uniforms were nicknamed 'robins' after the red-breasted bird. It is suggested that this association led to the robin being depicted on Christmas cards. Often the robin was depicted as carrying a card in its beak, with the bird becoming a postal messenger bearing Christmas greetings. The scarlet Post Office uniforms have long gone, but the robin is now firmly established as an iconic symbol of Christmas.

Reindeer

Nowadays, reindeer are very much part of how Christmas is depicted. Father Christmas (or Santa Claus), the gift giver, rides through the skies on a sleigh drawn by a team of reindeer. Then there is Rudolph the Red-Nosed Reindeer, a character created by Robert L. May in a 1939 story that a decade later was turned into a popular song by Johnny Marks. The association between reindeer and Santa Claus is thought to date back to the nineteenth century and is credited to an American writer called Clement Clarke Moore. In 1823, his poem *T'was the Night before Christmas*, also known as

A Visit from St Nicholas, was published in a newspaper and became remarkably popular. The poem featured St Nick (a precursor of Santa Claus) driving a sleigh drawn by reindeer, as seen in this extract:

> *The moon on the breast of the new-fallen snow,*
> *Gave the lustre of mid-day to objects below,*
> *When what to my wondering eyes should appear,*
> *But a miniature sleigh, and eight tiny reindeer,*
> *With a little old driver, so lively and quick,*
> *I knew in a moment it must be St Nick.*

Reindeer, also known as caribou, are members of the deer family. Their geographic distribution is circumpolar, and they live in Arctic tundra and subarctic (boreal) forest regions. They are specially adapted for life in harsh Arctic environments, their striking antlers being an important aspect of this. Unlike other deer, both male and female reindeer have antlers, and in proportion to their size these are among the largest of any deer. The antlers serve a practical purpose in allowing the reindeer to scrape away the snow and burrow through soil to find food such as lichens. Indeed, they are the only large mammal able to metabolise lichens owing to specialised bacteria in their gut. They also have two layers of fur (one ultrafine and dense, the other insulating), hooves that adapt to the seasons and eyes sensitive to ultraviolet light to enable them to cope with the long, dark Arctic winters.

Circumpolar people – among them the Sami in Sapmi (also known as Lapland) and the Inuit in North America – have developed a close relationship with the reindeer. Reindeer herding has been an important way of life for the nomadic Sami people for centuries. The reindeer are used for food, clothing and trade and also as draft animals. In-depth knowledge of both reindeer and their environment is apparent in the Sami language, which has a richly nuanced vocabulary for different types of snow: for example,

EUROPE.

Coloured lithograph
of reindeer.

the word *seanas* describes a dry, large-grained type of snow found in late winter and spring, which is easy for the reindeer to dig through.

Kew Gardens played a part in an intriguing story which saw the experimental introduction of reindeer from Sapmi to the people of Labrador in Canada, in the early years of the twentieth century. The medical missionary Wilfred Grenfell came up with the idea of introducing reindeer to provide these people with a sustainable source of meat, milk, clothing and draft animals, instead of them having to continue relying on harp seals for food and dogs for draft animals. In order to see if this would be viable, botanical research was done into the availability of lichens and mosses in Labrador for the reindeer to feed on. Kew was consulted, and it was determined that at least four lichen species found in the area would serve as a food source for reindeer. Three hundred reindeer and three Sami herders were introduced to Labrador, and initially the project was considered a success. Ten years later, however, the herd had reduced in size and the remaining reindeer were transported away from the area. In an unfortunate twist, unknown to Grenfell, the herd were carrying a parasite which affects reindeer in Newfoundland to this day.

The ox

Cattle, domesticated around 10,500 years ago, have long been valued as draft animals and for their meat, milk, leather and horns. The ox, a sacrificial animal, came to symbolise a number of attributes including strength, perseverance, dependability and endurance. In the Christmas story, an ox is part of the Nativity scene, frequently depicted in images of Mary, Joseph and the baby

Nativity by Master IQV, c. 1543.

Jesus in the stable at Bethlehem. St Francis of Assisi – noted for his love of animals – is said to have created the first-ever Nativity scene in 1223, with a living Christmas crib set up near Greccio in Italy, complete with a manger and a live ass and a live ox. Recreating the Christmas crib or Nativity scene became a popular custom in western Europe.

One piece of folklore tells how the cattle kneel on Christmas Eve in remembrance of the birth of Christ. Hearing this story from his mother, the English author Thomas Hardy refers to it in his novel *Tess of the D'Urbervilles* (1891). He also memorably evokes the legend in his 1915 poem entitled *The Oxen*, which is set on Christmas Eve.

Christmas Eve, and twelve of the clock.
'Now they are all on their knees,'
An elder said as we sat in a flock
By the embers in hearthside ease.

We pictured the meek mild creatures where
They dwelt in their strawy pen,
Nor did it occur to one of us there
To doubt they were kneeling then.

So fair a fancy few would weave
In these years! Yet, I feel,
If someone said on Christmas Eve,
'Come; see the oxen kneel,

'In the lonely barton by yonder coomb
Our childhood used to know,'
I should go with him in the gloom,
Hoping it might be so.

Snow

The idea of a 'white Christmas' has become very much part of how we like to picture Christmas. So much so that each year, bookies in the UK take bets on its likelihood. Christmas is a festival filled with nostalgia. That sense of looking wistfully back to an ideal Christmas is apparent in Irving Berlin's lyrics to his famous song *White Christmas* (1942), which begins like this:

> *I'm dreaming of a white Christmas*
> *Just like the ones I used to know*
> *Where the tree tops glisten*
> *And children listen*
> *To hear sleigh bells in the snow, oh, the snow*

Christmas in the northern hemisphere is a winter festival, so its association with snow and ice is not totally surprising. In Jane Austen's novel *Emma*, published in 1816, a downfall of snow is regarded as a natural aspect of the festive season: '"Christmas weather," observed Mr Elton. "Quite seasonable."' Charles Dickens grew up experiencing bitterly cold, snowy winters. In his *A Christmas Carol*, the Ghost of Christmas Present transports Ebenezer Scrooge to Christmas Day in London, where 'The people made a rough, but brisk and not unpleasant kind of music, in scraping the snow from the pavement in front of their dwellings, and from the tops of their houses.' Snow also plays a part in Washington Irving's *Old Christmas* (1876), his nostalgic depiction of a traditional Christmas: 'But in the depth of winter, when nature lies despoiled of every charm, and wrapped in her shroud of sheeted snow, we turn for our gratifications to moral sources.'

With the coming of winter and the lowering temperatures, most of the outdoor plants at Kew Gardens slow down in order to

Der Schlitten.

Coloured lithograph of sledging in the snow.

preserve their energy. An example of the way some plants have adapted to snowy conditions can be found in the Rock Garden, which was constructed in 1882 and is more than an acre (4,000 m^2) in size, making it one of the oldest and largest of its kind in the world. It is a vital resource for Kew's research. Around 60 per cent of plants displayed here are grown from wild-collected seed. Set within the Rock Garden, the Davies Alpine House showcases a variety of alpine plants. Striking in appearance, this glasshouse is designed to recreate the dry, light, cool and windy conditions that alpines need to thrive, but without using energy-intensive air conditioning and wind fans. Conceived as two back-to-back arches, the structure creates a stack effect that draws warm air out of the building. Below ground, air is cooled passing through a concrete labyrinth and recirculated around the perimeter, while the low-iron glass exterior allows 90 per cent of the light to pass through.

The alpine plants growing in the Rock Garden are among some of the most resilient in the world, able to grow above the altitude at which trees survive. In the wild, they spend the winter dormant under a blanket of snow. The snow, in fact, has a protective effect. While it is freezing under the snow, conditions won't reach the extreme cold of the surface. The lack of light under the snow is not a problem as the plants are sleeping, not growing at all. Plants that live in freezing conditions have compounds in their cells that lower the freezing point of the cells' contents, acting like antifreeze in a car or salt on a road, and prevent ice crystals from forming in the cells. They grow close to the ground, often as mats or cushions of foliage, so they are less exposed to icy, drying winds. Their leaves often have a thick, waxy cuticle that reduces moisture loss and provides added protection from the cold. The melting snow in spring provides moisture and exposes the plants to light. They have to flower and set seeds quickly in order to make the most of the short growing season.

Frankincense and myrhh

Gold, frankincense and myrrh: of the three precious gifts that the Magi brought to offer to the baby Jesus in Bethlehem, it is striking that two of them are plant-based. Both frankincense and myrrh are resins made from dried tree sap. Frankincense comes from trees belonging to the genus *Boswellia*, while myrrh comes from trees belonging to the genus *Commiphora*. The resins were – and, indeed, still are – highly valued for their fragrant properties and burnt as incense or used in anointing oils. The word 'frankincense' derives from the Old French *franc encense*, meaning 'incense of high quality'. The sap is obtained by tapping the trees and scraping out the oozing sap, which then hardens into resin.

Frankincense has long been bought and sold as a commercial commodity, with its trade spanning more than five thousand years. There is a fascinating account of how it was gathered by Pliny the Elder in his *Natural History*:

> *The first, and what we may call the natural, vintage, takes place about the rising of the Dog-star, a period when the heat is most intense; on which occasion they cut the tree where the bark appears to be the fullest of juice, and extremely thin, from being distended to the greatest extent. The incision thus made is gradually extended, but nothing is removed; the consequence of which is, that an unctuous foam oozes forth, which gradually coagulates and thickens. When the nature of the locality requires it, this juice is received upon mats of palm-leaves, though in some places the space around the tree is made hard by being well rammed down for the purpose. The frankincense that is gathered after the former method, is in the purest state, though that which falls on the ground is the heaviest in weight: that which adheres to*

Frankincense (*Boswellia sacra*) from *Köhler's Medizinal Pflanzen*, Franz Eugen Köhler, 1887.

Balsamodendron Mÿrrha Nees v. Es.

the tree is pared off with an iron instrument, which accounts for its being found mingled with pieces of bark.

The Economic Botany Collection at Kew Gardens holds 140 frankincense specimens. One of these is a lump of resin collected by the botanist Isaac Bayley Balfour on an 1880 expedition to the island of Socotra in the Indian Ocean. The island has many plants not found elsewhere, and Balfour assigned the botanical name *Boswellia ameero* to the tree from which the resin came. Because there are about thirty species of *Boswellia* (many of which produce resins that have been traded over great distances), there has long been much confusion about frankincense. Research by plant taxonomists at Kew and elsewhere has helped cast light on its complex story. With regard to the frankincense brought by the Magi, it is now clear that the species that grow in the vicinity of the Near East are found in two areas: in the part of Africa south of Egypt, *Boswellia frereana* and *B. sacra* grow in Somalia, while *B. papyrifera* grows in Ethiopia and Sudan; in southern Arabia, in Oman and Yemen, *B. sacra* is the source of frankincense. Today, *B. papyrifera* is the most important species in international trade.

One important role of the Economic Botany Collection is to supply reference samples to researchers. Modern methods of chemical analysis require only tiny amounts of resin (around two hundredths of a gram), so where there is good reason, destructive sampling is possible.

Myrrh, which like frankincense was obtained by tapping trees for their sap, was also greatly prized for its aromatic fragrance. It was used in embalming by the ancient Egyptians and for making incense in India. It is mentioned a number of times in the Bible, where it is an ingredient in a holy anointing oil for priests and purification rituals. After the Crucifixion, Joseph of Arimathea brings myrrh and aloe with which to embalm the body of Jesus.

Unfortunately, nowadays, the overharvesting of both

Myrrh (*Commiphora myrrha*) from *Köhler's Medizinal Pflanzen*, Franz Eugen Köhler, 1887.

frankincense and myrrh is threatening the survival of these wild trees. Traditionally in Somaliland, the harvesting of frankincense was done for only six months of the year, with the tree allowed a year to rest. Economic pressures mean that the trees are now harvested every year. Without a respite period, the trees cannot repair themselves and their immunity is weakened, leading to mortality. There are also concerns that excessive harvesting of young trees – before they have reached full, productive size – will affect the trees' fertility and have a damaging impact on their numbers.

The goose

For centuries, a central element of Christmas dining was roast goose, as well as other roast meats such as beef or pork. Before the arrival of the turkey, the goose was the largest of the domestic birds in the UK, which meant it lent itself to being served at a feast. Another practical reason for its festive association was that goose was historically a bird that was eaten in the autumn and winter months, having been fattened up on the post-harvest stubble found in the fields. There is a long tradition of dining on goose on September 29th to celebrate Michaelmas, the Feast of St Michael the Archangel – it was believed that eating a goose on that day would bring prosperity. One apocryphal legend tells that Queen Elizabeth I was dining on roast goose on Michaelmas when she heard the Spanish Armada had been sunk and proclaimed that geese should be eaten to commemorate the defeat.

Roasting meat or poultry over a fire, which requires extravagant amounts of fuel and labour, has been a highly prized form of cooking since medieval times. In *A Christmas Carol*, Dickens makes it clear that the Cratchits can't afford to roast their goose at home, instead going to collect it from the baker's, where it's been roasted for them. 'Such a bustle ensued that you might have thought a goose the rarest of all birds,' writes Dickens, 'a feathered phenomenon, to which a black swan was a matter of course – and in truth it was something very like it in that house.'

Geese were farmed in the east of England, notably the arable regions of Norfolk and Lincolnshire. For hundreds of years, annual goose fairs were a feature of the British agricultural calendar. The Nottingham Goose Fair was especially famous, with the first known reference to a goose fair – dating from 1541, and calling it 'Gose feyre dey' – pertaining to this particular one. The story goes that geese were driven from the Lincolnshire fens to be sold at this fair. Today's Nottingham Goose Fair is held in October and is a large, popular funfair.

Nowadays, goose is distinctly a luxury, in recent years achieving a niche following as a Christmas bird, with diners appreciating its rich, moist, flavourful meat. Roasting a goose results in copious amount of fat coming off during its time in the oven. This should be reserved and stored in the fridge. Goose fat, which was once also known as goose grease because of its softness, has long been prized. One excellent use for goose fat is in cooking roast potatoes, a popular accompaniment to a festive roast bird or joint of meat. Aromatic sage (*Salvia officinalis*) has long been favoured in the UK as an accompaniment for goose. The bestselling eighteenth-century cookbook author Hannah Glasse, in her 'Directions for roasting a goose', advises that sage leaves rolled in butter be placed in its cavity. In the same recipe, however, she warns, 'Never put onion into any thing unless you are sure that every body loves it.' Despite Glasse's reservations, however, sage and onion has become the classic stuffing for roast goose.

Flock of Geese by Elizabeth Nourse, c. 1883.

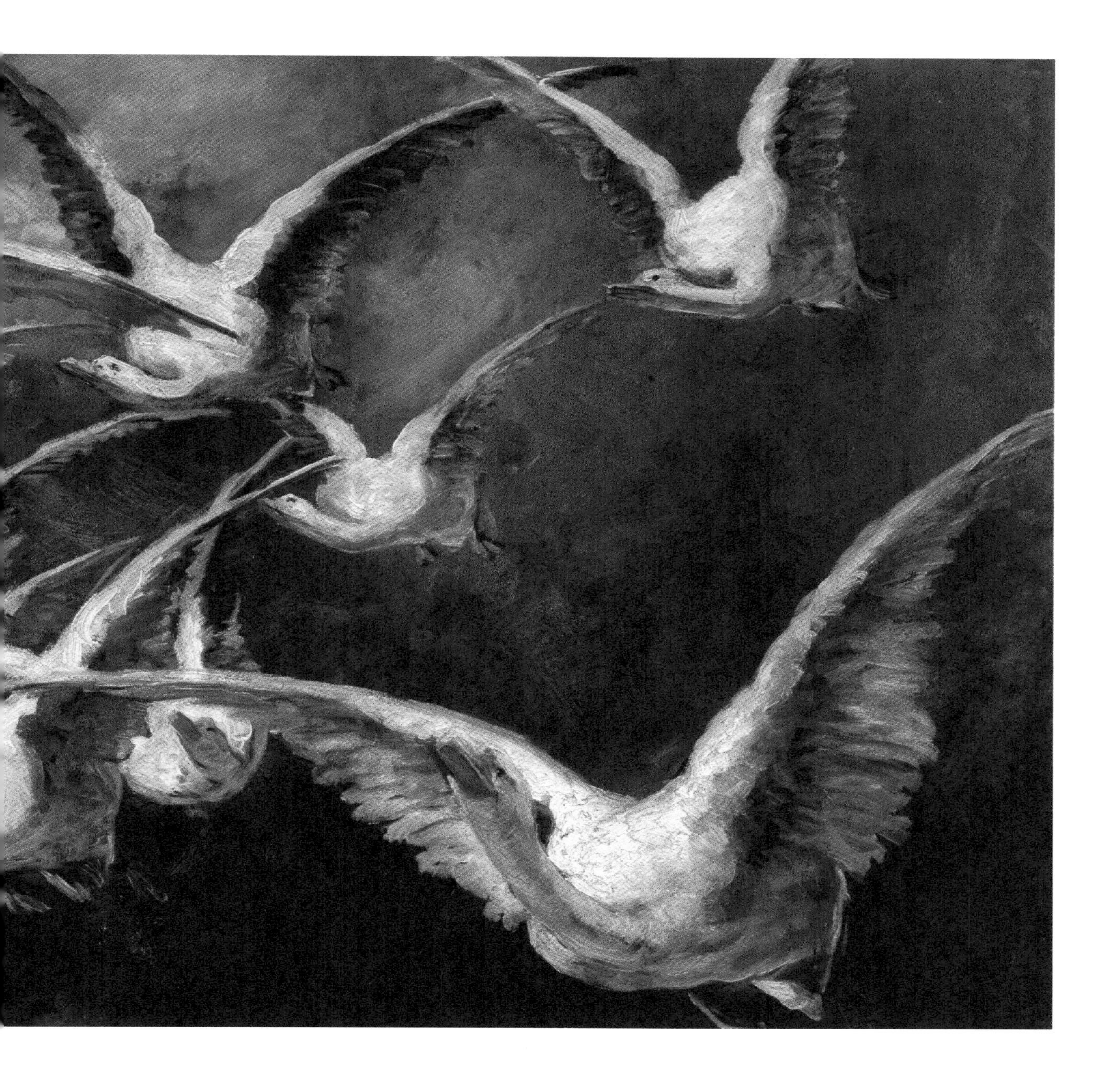

The turkey

Today, a roast turkey has become **the** Christmas bird in the UK, with its presence on the Christmas table almost obligatory. Originally from Central America, the turkey was introduced into Europe by the Spanish conquistadors. In 1511, King Ferdinand of Spain ordered that every ship sailing from the Indies to Spain should bring five breeding pairs of turkeys on it, and thus a taste for turkey spread through Europe. Trading routes, however, led to a perception that the bird had come originally from Turkey, hence its name. The turkey's large size, which meant it could feed a number of people, together with its tender meat, saw it become a bird eaten at feasts, gradually displacing other large birds such as bustards, swans, herons and peacocks. By 1573, when the farmer and poet Thomas Tusser wrote his instructional poem *Five Hundred Points of Good Husbandry*, the turkey was considered a Christmas food. In Tusser's description of Christmas cheer, he writes of:

Beef, mutton, and pork, shred pies of the best,
Pig, veal, goose, and capon, and turkey well drest,
Cheese, apples, and nuts, jolly Carols to hear,
As then in the country, is counted good cheer.

In East Anglia, with its cereal crops, turkeys were fattened on the stubble left in the fields after the annual harvest. The area became noted for its turkeys, the Norfolk black becoming a popular breed. English writer Daniel Defoe, in his 1722 *Tour of the Eastern Counties of England*, observed: 'This county of Suffolk is particularly famous for furnishing the city of London, and all the counties round, with turkeys; and tis thought, there are more turkeys bred in this county, and the part of Norfolk that adjoins to it, than in all the rest of England.' Large flocks of turkeys were driven to the London

Turkey print by R. Schulz, 1829–1880.

markets on foot, with the birds' feet reportedly wrapped in leather or dipped in tar so that they could walk along the rough roads.

The turkey's association with celebratory meals continued to grow. By 1786, 'alderman' had become a slang term for a roast turkey garnished with a chain of sausages, a reference to the gold chain worn by aldermen (municipal officials who were often satirically depicted as plump and prosperous). The turkey's presence on the Christmas table is cemented by its special role in Dickens's *A Christmas Carol*. Towards the end of the story, when Scrooge, the reformed miser, realises it is Christmas Day, his first act is to lean out of a window, call to a boy and send him on an errand to buy the 'prize turkey' from a nearby poulterers. When the boy returns with the bird, its size is made clear in an emphatic image: 'It was a Turkey! He never could have stood upon his legs, that bird. He would have snapped 'em short off in a minute, like sticks of sealing wax.' It is this magnificent bird which Scrooge sends to the Cratchit family as a generous gift.

The boar's head

One striking, medieval banquet dish long associated with Christmas was the boar's head. Wild boar was a prestigious meat, and cooking the head of the boar was a long, complex procedure involving debristling, brining, stuffing and poaching. It's a sign of how highly regarded the dish was that it has its own Christmas carol, first published in 1521, which is sung when the boar's head is presented at a feast, classically carried in on a platter decorated with bay and rosemary.

The boar's head in hand bear I,
Bedecked with bays and rosemary;
And I pray you, my masters, be merry,
Quot estis in convivio.

Caput apri defero,
Reddens laudes Domino.

The boar's head, as I understand,
Is the bravest dish in all the land;
When thus bedecked with a gay garland,
Let us servire cantico.

Caput apri defero,
Reddens laudes Domino.

Our steward hath provided this,
In honour of the King of Bliss;
Which on this day to be served is
In Reginensi Atrio.

Caput apri defero,
Reddens laudes Domino.

The stuffed boar's head – so evocative of medieval times – has lasted down the centuries as a special dish served at feasts. Wild boar became extinct in England by the seventeenth century, so pigs' heads were often used instead (wild boar were eventually reintroduced to the country in the late 20th century). A stuffed boar's head was one of a number of splendid dishes served to Queen Victoria on Christmas Day, the head having been sent as a gift usually from either the emperor of Germany or the king of Saxony.

In London, the Worshipful Company of Butchers still holds the annual Boar's Head Ceremony in February, the origins of which date back to 1343. That year, the Lord Mayor of London decreed that a piece of land be made available to the city's butchers where they could deal with their animal waste products. This land was next to the River Fleet, in which they were granted permission to dispose of the waste. The rent the butchers had to pay was in the

Wild boar print by R. Schulz, 1879.

form of a boar's head. Over time, this evolved into a procession bearing a boar's head from the Butchers' Hall to Mansion House. These days a model of a boar's head is carried in the procession, but a real one awaits the participants at Mansion House, where they enjoy a reception.

In Oxford, The Queen's College holds an annual boar's head dinner in December. Legend has it that a student from the college was walking through the forest of Shotover when he was attacked by a wild boar. With great presence of mind, he rammed a volume of Aristotle into the animal's mouth and choked it to death, so the dinner commemorates his escape. A nineteenth-century account of the meal describes it as follows:

> *Suffice it to mention that on Christmas Day a large boar's head, adorned with a crown, wreathed with gilded sprays of laurel and bay, as well as with mistletoe and rosemary, and stuck all over with little banners, is solemnly carried into the Hall by three bearers. A flourish from a trumpet announces the entry. The bearers are accompanied by a herald, who sings the old English Song of the Boar's Head. At the end of each verse those present join in the Latin refrain.*

The so-called Boar's Head Gaudy continues to be held at The Queen's College each year, though rather than taking place on Christmas Day, it is now held on the nearest Saturday before Christmas, or if that falls on Christmas Eve, on the Saturday before. The singing of the *Boar's Head Carol* as the animal's head is brought in remains part of the ceremony.

Brassicas

The winter months in northern Europe see seasonal members of the Brassicaceae family come into their own. Known also as cruciferous vegetables, this large family includes broccoli, cabbages, cauliflower and kale. They are rich in nutrients and minerals and contain natural compounds known as glucosinolates, which may have possible health benefits.

Members of the family play their part in festive European cooking. Red cabbage – which is, in fact, a striking purple colour rather than red – gains its colour from the presence of pigments known as anthocyanins. In Germany, roast goose is the traditional centrepiece of the Christmas table, a popular accompaniment being braised red cabbage cooked with apple and vinegar, so that its sharpness contrasts nicely with the richness of the meat. In Denmark, similarly braised red cabbage is served as part of the traditional Christmas Eve dinner with roast pork or duck. In eastern Europe, sauerkraut – fermented cabbage – is a popular element in Christmas dishes: it's used in soups, hearty meaty stews and the Polish dumplings known as pierogi.

In the UK, one brassica above all others has become especially associated with Christmas– namely, the Brussels sprout (*Brassica oleracea*). Striking in appearance, these small green balls – which look like tiny, tightly furled cabbages – are edible buds which grow on long, thick stalks. Indeed, its cultivar group name Gemmifera means bud-producing. Nowadays, it is possible for consumers to buy Brussels sprouts both loose and on stalks. Little is known for certain about where this vegetable was first cultivated, though the distinctive name suggests a link to Belgium. The nineteenth-century English food writer Eliza Acton gives a recipe for Brussels sprouts in the 'Belgian mode' in her popular 1845 cookbook *Modern Cookery for Private Families*. She observes, 'These delicate little sprouts or miniature cabbages, which at their fullest growth scarcely exceed a large walnut in size, should be quite freshly gathered.' This recipe sees the sprouts boiled until tender, drained well and served on toasted bread buttered on both sides with extra melted butter on the side. It is unclear why Brussels sprouts became a traditional element of the Christmas meal, though their seasonal availability during the winter months would be a practical reason.

Intriguingly, Brussels sprouts, cabbage, broccoli and cauliflower all belong to one plant species: *Brassica oleracea*. Their wild ancestor is a spindly plant that grows on limestone rocks of the coastal Mediterranean region. Over the centuries, farmers domesticated this plant and created a number of cultivars. These

Brussels sprouts (*Brassica oleracea*) from *Plantarum Medicinalium*, Joseph Jacob Plenck Icones, 1794.

cultivars are marked by different attributes, such as the large buds of the Brussels sprout, the sizeable leaves of cabbage and kale and the densely packed flowers and thickened stems of broccoli and cauliflower.

There is no denying that the Brussels sprout has a divisive reputation as a food, being both loved and loathed. One reason for the bad reputation might be that when they're overcooked, they release a sulphurous odour. This is because, like other cruciferous vegetables, they are high in chemical compounds, which when exposed to heat for a period of time produce hydrogen sulphide. Contemporary ways of cooking the vegetable include stir-frying, roasting and charring. Flavourings and garnishes include chestnuts, bacon or pancetta, garlic, chillies, Parmesan cheese and pomegranate seeds. For many, the Christmas meal would not be complete without a bowlful of Brussels sprouts. Boxing Day bubble and squeak, made from roast potatoes and sprouts, is a satisfying way of using up the leftovers.

Christmas pudding

One of the traditional elements of a British Christmas Day lunch is the Christmas pudding, served at the end of the meal. There is a satisfying ritual element of drama to its serving – the lights are dimmed, brandy is heated through, poured over it and set alight, so that the rotund pudding arrives with blue flames dancing around it. Its accompaniments provide an added layer of indulgence: boozy butters flavoured with brandy, rum or Cointreau, rum sauce, cream or custard. In *A Christmas Carol*, Dickens depicts the pudding as a key element of the Cratchit family's festive meal.

> *In half a minute Mrs Cratchit entered: flushed, but smiling proudly: with the pudding, like a speckled cannon-ball, so hard and firm, blazing in half of half-a-quartern of ignited brandy, and bedight with Christmas holly stuck into the top.*
>
> *Oh, a wonderful pudding! Bob Cratchit said, and calmly too, that he regarded it as the greatest success achieved by Mrs Cratchit since their marriage.*

The ingredients that go into a Christmas pudding – dried fruit and nuts, fragrant spices – have a medieval flavour to them. Historically, though, long before Christmas pudding there was plum pottage – a rich, soupy mixture of spiced meat stock with dried fruit (the 'plum' of the title). This dish was enjoyed at feasts, and eventually became associated with Christmas. Hannah Glasse,

Advert for Borwick's Baking Powder, c. 1920s.

SIX GOLD MEDALS
SIX GOLD MEDALS
BORWICK'S
BAKING
POWDER
IS THE
BEST

Trade card from the *New Years 1890 Cards* series, Kinney Brothers Tobacco Company, 1889–90.

the author of the eighteenth-century bestselling cookbook *The Art of Cookery Made Plain and Easy* (1747), gives a recipe for a Christmas plum porridge made from beef stock thickened with bread and enriched with spices, dried fruit and alcohol. By this time, however, the plum pudding had become popular, made using suet and dried fruit such as raisins and currants. Glasse's cookbook also includes a recipe for a boiled plum pudding, flavoured with nutmeg and ginger and boiled for five hours. The invention of the pudding cloth – rather than the animal guts used historically for savoury puddings – is seen as playing a key part in the development of the sweet suet pudding.

The Christmas pudding (named as such) became popular in the late nineteenth century. The tradition of adding small tokens – silver coins, buttons, thimbles or specially made charms – to the pudding mixture to bring good fortune to their finders adds to its sense of being a special food. This practice of adding a lucky token to a celebratory dish is one that is found in many parts of Europe, such as in France, where the galette des rois contains a *fève* (bean). Stir Up Sunday – the Sunday before Advent, the day when the Collect read in Church of England churches begins 'Stir up, we beseech thee' – became the day for making your Christmas pudding at home. From a practical point of view, the fact that suet puddings keep well and are transportable was a factor in the rise of the Christmas pudding. Between 1847 and 1851, Joseph Hooker, who later became director of Kew, travelled to India and the Himalayas on a plant collecting expedition. In one of his letters home, he wrote of his Christmas dinner aboard the steam frigate Moozuffer, 'Of Roast Beef we have none; but the more easily compassed Plum pudding was present.'

Nowadays, commercially produced Christmas puddings are widely available. However, in many families, the tradition of having homemade Christmas pudding that's matured in advance of Christmas Day continues to be lovingly maintained.

Festive spices

Spices are a reminder of just how versatile and varied culinary plants are. Largely speaking, spices consist of seeds, buds, herbs, roots and fruits, usually but not always dried. They play a special part in our cooking, enhancing our food by adding flavour and fragrance to it. For many centuries, spices were hugely prized, appreciated for their medicinal properties as well as their organoleptic ones. As a result of their value, the history of spices is, in many cases, a dark and bloody one. Many spices are native to tropical countries. European countries colonised the places where precious spices came from in order to create and retain monopolies

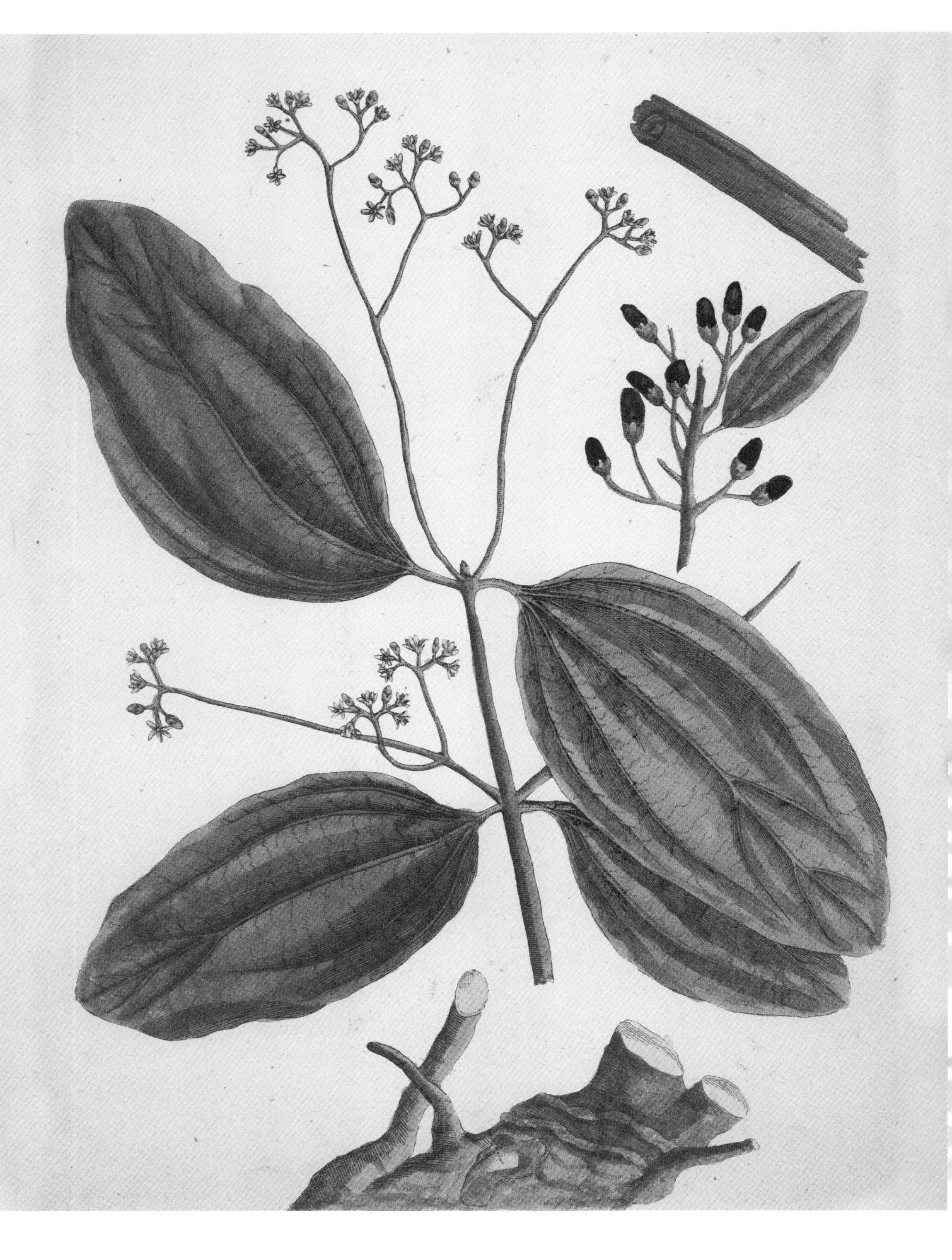

on what was a vastly lucrative trade. Cinnamon, cloves and nutmeg are all historic examples of this practice.

Today, spices are affordable and widely available, but they still retain a special place in our kitchens. Over the centuries, certain spices in particular have become especially associated with Christmas, with their scents an evocative part of this celebratory season.

CINNAMON

Sweetly fragrant cinnamon, which is the prepared bark of the evergreen tree *Cinnamomum verum*, was for centuries a highly prized spice with much folklore associated with it. The ancient Greek historian Herodotus, writing in his *Histories* in around 430 BCE, speculated on its mysterious origins. He recorded the remarkable tales told by the Arab traders who dealt in cinnamon:

> *Where the wood grows, and what country produces it, they cannot tell – only some, following probability, relate that it comes from the country in which Bacchus was brought up. Great birds, they say, bring the sticks which we Greeks, taking the word from the Phoenicians, call cinnamon, and carry them up into the air to make their nests.*

In reality, Sri Lanka is the home of true cinnamon and the spice is a major export for the country. Cinnamon production is laborious and time-consuming. Shoots from the coppiced cinnamon tree are cut and processed immediately after harvesting. The inner bark is loosened, pried off and dried. It is used both in its tightly rolled form (known as quills) and as a ground powder. Cinnamon is widely used at Christmastime, adding aroma to mulled wine and as a flavouring in Christmas mincemeat and German *lebkuchen*.

Cassia, also known as Chinese cinnamon, is another aromatic bark with a similar aroma to that of true cinnamon. It comes from

Cinnamon (*Cinnamomum verum*) from *A Curious Herbal*, Elizabeth Blackwell, 1739.

Cinnamomum burmanni, another tree in the laurel family. Cassia is thicker and coarser than cinnamon and has a more pungent and pronounced aroma and flavour. It is used in Chinese cooking to flavour meat dishes and as one of the spices in Chinese five spice powder.

Clove (*Syzygium aromaticum*) from *Curtis's Botanical Magazine*, 1827

CLOVES

The dried, unopened buds of the clove tree (*Syzygium aromaticum*) form the spice we know as cloves. Their Italian name, *chiodi di garofano* (literally 'nails of carnation'), is expressive both of their appearance and their powerful aroma. Sturdy, pointed cloves are used to stud Christmas hams and also the onion used to infuse the milk for a classic bread sauce. Their flavour is so assertive that they should be used judiciously, since they can easily dominate a dish. Cloves have a numbing effect, owing to the fact that they contain eugenol, which is an anaesthetic. Eugenol also has antiseptic properties, and oil of cloves was once used as a natural remedy to treat conditions such as toothache.

The Marianne North Gallery at Kew Gardens displays paintings of a number of spices by the redoubtable Victorian botanical artist after whom it's named. Among them are pictures of both cloves and cinnamon painted while she was in the Seychelles in 1883. She wrote in her autobiography *Recollections of a Happy Life* (1893):

> *The trees of both cloves and cinnamon were from twelve to twenty feet high, and every leaf and twig was sweet, the young leaves of the most delicate pink colour.*

GINGER

Ginger – with its distinctive aromatic warmth – plays a central role in festive baking (see Gingerbread, page 104). It is the underground

Ginger (*Zingiber officinale*) from *Köhler's Medizinal-Pflanzen*, Franz Eugen Köhler, 1887.

stem (rhizome) of the ginger plant (*Zingiber officinale*), whose native range is from India to south-central China. Ginger has long been valued for its medical and culinary uses in China, India and South East Asia.

A much-traded spice, it was highly regarded in ancient times by the Assyrians, Babylonians, Egyptians, Greeks and Romans. Although ginger can be used fresh or dried, it was in the dried form in which it was transported that it was predominantly used in Middle Eastern and European cuisine. By the early seventeenth century, Europeans had developed a taste for stem ginger preserved in syrup, an ingredient which is today often used in ginger cakes, puddings and biscuits.

NUTMEG

From the fruit of the nutmeg tree (*Myristica fragrans*) we get the large, light-brown seeds we know as nutmeg as well as mace, which is made from the lacy aril (seed covering) that encases the seed. Nutmeg's rich, warm aroma is an element in a number of classic festive treats, such as Christmas pudding, Christmas mincemeat, Christmas fruit cake, eggnog and punch.

The nutmeg tree was indigenous to the Banda Islands (known as the Spice Islands) in Indonesia, with the Dutch long holding a monopoly on the nutmeg trade. British attempts to break this monopoly saw nutmeg trees being introduced to other tropical countries. The eighteenth-century naturalist Sir Joseph Banks, Kew Gardens' first unofficial director, introduced the nutmeg tree to Grenada. The spice remains an important export for the Caribbean island to this day.

The fruit of the nutmeg tree is harvested when ripe, and the flesh and mace (bright red when fresh) are stripped off. The nutmeg kernels are retained inside their shells for a number of weeks to dry until they rattle. The shells are then cracked and the smooth, brown kernels are removed and sold either whole or finely ground. It is sometimes possible to buy nutmegs inside their shells with their mace coating still intact.

Nutmeg (*Myristica fragrans*) from *Köhler's Medizinal-Pflanzen*, Franz Eugen Köhler, 1890.

Nutmeg is best used freshly grated. In the seventeenth and eighteenth centuries, small, portable nutmeg graters made from silver became a fashionable item to own. These were carried in pockets or in travelling canteens, alongside cutlery, a beaker and a corkscrew. Popular shapes included hearts and shells, as well as cylindrical and oval boxes. By the nineteenth century, nutmeg graters had become far more mundane, everyday objects. The Victorian journalist Henry Mayhew in his *London Labour and the London Poor* (1851) describes a 'crippled street-seller of nutmeg graters' whose wares were made of tin and usually cost 'only a penny a piece'.

Charles Dickens, who was fond of punch, had a silver nutmeg box. Nutmeg graters – written about with characteristic, creative energy – appear in two of his novels. In *David Copperfield* (1850), David remembers Peggotty (his nurse and the family housekeeper) as follows:

> *I have an impression on my mind which I cannot distinguish from actual remembrance, of the touch of Peggotty's forefinger as she used to hold it out to me, and of its being roughened by needlework, like a pocket nutmeg-grater.*

In *Great Expectations* (1861), the narrator, Pip, muses on the appearance of his sister Mrs Joe:

> *My sister, Mrs Joe, with black hair and eyes, had such a prevailing redness of skin that I sometimes used to wonder whether it was possible she washed herself with a nutmeg-grater instead of soap.*

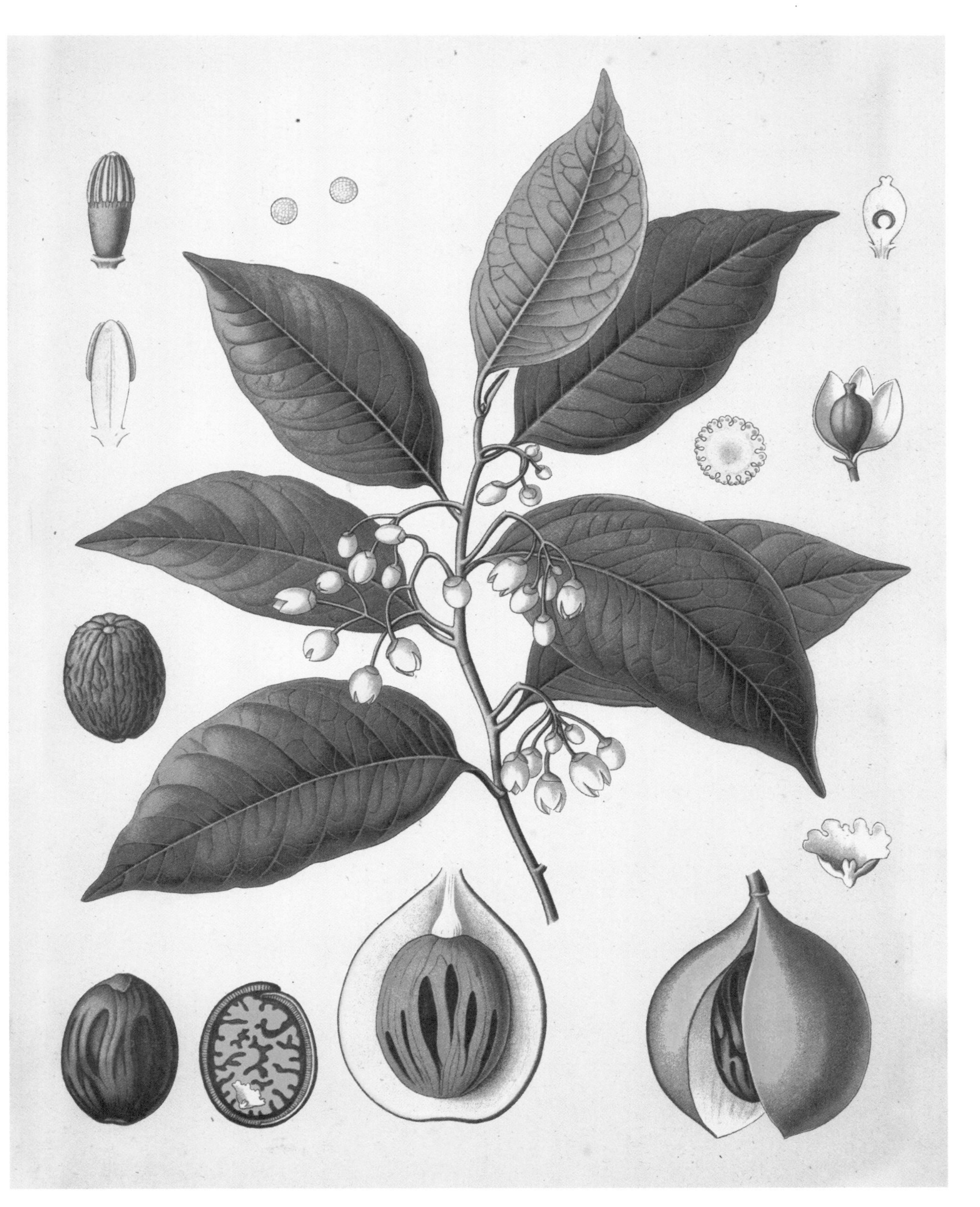

The almond

Among the plant-derived foods that we enjoy eating during the Christmas season are a variety of nuts. We serve salted, roasted nuts as a nibble with drinks, place bowls of nuts in their shells with a nutcracker on a table to enjoy in a leisurely way after a meal, and use nuts widely in baked goods such as biscuits and desserts. One nut that plays a special part at Christmas owing to its use in marzipan (see page 91), *turrón*, *torrone* and nougat, is the almond, produced by the almond tree (*Prunus amygdalus*). Botanically speaking, the almond is not considered a true nut. A nut is a dry fruit which consists of a hard shell covering a single seed. Chestnuts are an example of a true nut. Almonds, however, are drupes or stone fruit, a fleshy fruit with a thin skin and a central stone containing a seed. Whereas many drupes (such as peaches) are valued for their fruit, almonds are valued for their seeds and considered to be a nut, culinarily speaking.

Of the tree nuts grown around the world, almonds form the largest crop. Owing to their low levels of polyunsaturated fats and high antioxidant vitamin E content, they keep relatively well. Across many cultures, almonds have historically had a symbolic value, representing qualities such as purity, watchfulness and fertility. To this day, sugared almonds are often given as wedding favours to guests.

In Europe (notably Spain, Italy and France), almonds are widely used to make nougat, a confection historically made from egg whites, honey and roasted nuts and thought to have originated in the Middle East, traditionally enjoyed over the Christmas period. In Spain, nougat (called *turrón*) has existed for centuries, and was very much relished as a courtly treat. Certain places in Spain are particularly known for their *turrón*, the version from the province

Almond (*Prunus amygdalus*) from *Köhler's Medizinal-Pflanzen*, Franz Eugen Köhler, 1887.

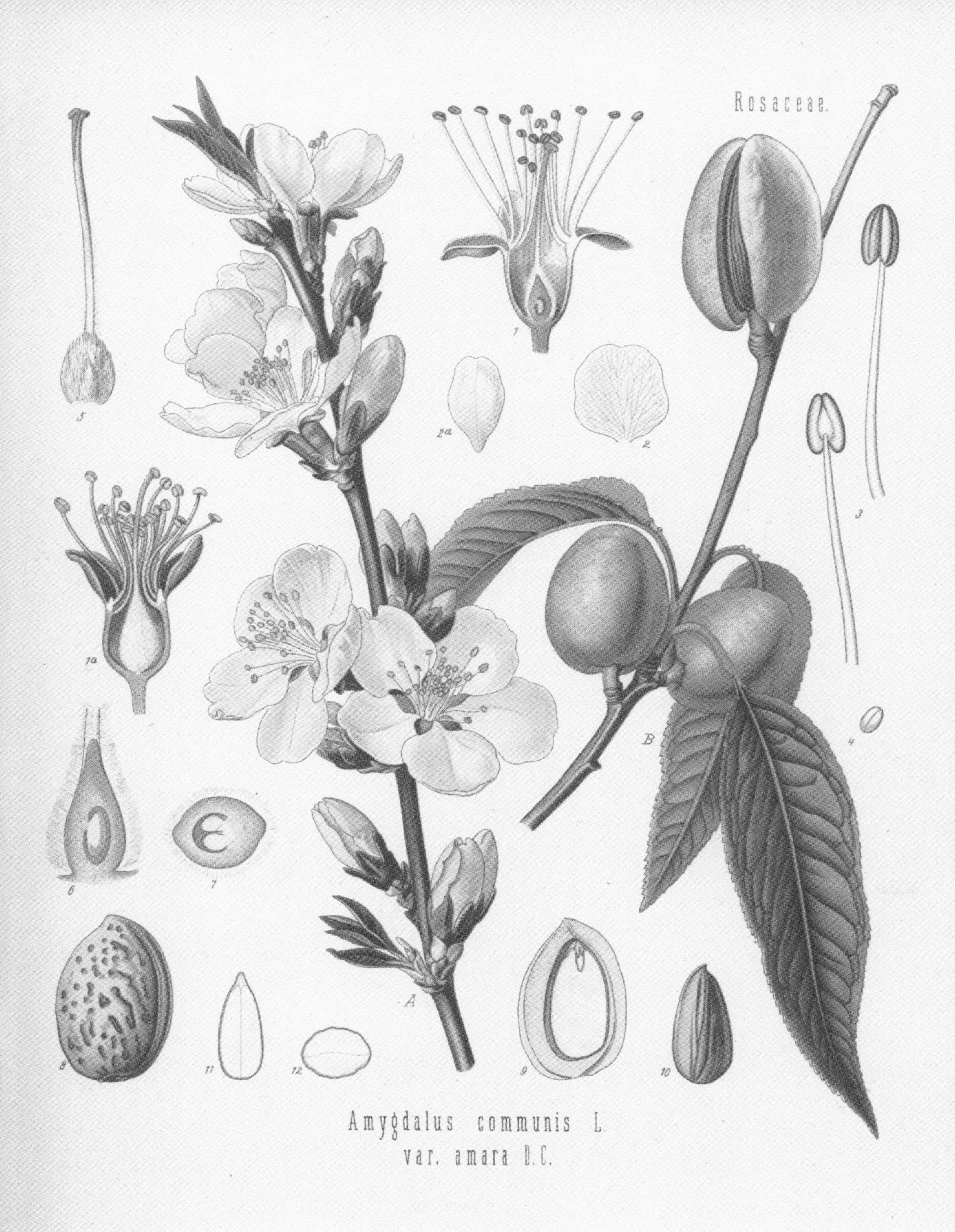

Amygdalus communis L.
var. amara D.C.

of Alicante being a famous hard, white nougat made using the high-quality Marcona almond, valued for its sweet flavour. For centuries, the city of Jijona in Alicante has been noted for its own *turrón*, a distinctive, soft-textured nougat made from ground almonds.

Italy has a number of places associated with its nougat (called *torrone*), including Cremona in Lombardy and Bagnara in Calabria, the latter's version distinctively flavoured with cocoa powder and spices. In France, the Provençal town of Montélimar has a long tradition of nougat production, linked to the introduction of almond trees to the region by French agronomist Olivier de Serres (1539–1619). By legal definition, *nougat de Montélimar*, which is a soft nougat, must be made from certain proportions of almonds, lavender honey and pistachios.

Such is nougat's historic significance that *turrón de Alicante*, *turrón de Jijona*, *torrone di Bagnara* and *nougat de Montélimar* have all been granted PGI (Protected Geographical Indication) status, which lays down in law where and how they are made. In France, Italy and Spain, nougat has long been established as a Christmas treat; for example, it is considered a classic component (along with dried fruits, nuts and special baked goods) of the Provençal tradition of *les treize desserts* – the '13 desserts' eaten on Christmas Eve.

The chestnut

The sweet chestnut tree (*Castanea sativa*) has long been valued for its edible nuts, found encased inside prickly green seed cases. Native to southern Europe and Asia Minor, it is believed to have been introduced into the UK as a food source by the Romans.

Sweet chestnut (*Castanea sativa*) from *Flora von Deutschland, Osterreich und der Schweiz*, Otto Wilhelm Thomé, 1886–9.

A member of the same family as the oak and the beech, it is a deciduous tree that can live for hundreds of years. One ancient example is the huge Chestnut of the Hundred Horses found on Mount Etna in Sicily, believed to be between two thousand and four thousand years old. Kew Garden's oldest trees are sweet chestnuts, thought to have been planted in the early eighteenth century.

Chestnuts are higher in starch and lower in oil than other nuts. This means that they can be dried and ground into a flour, a useful attribute which the Romans valued. To this day, chestnut flour is used as an ingredient in Italian cooking, featuring in dishes such as Tuscany's *castagnaccio* (a flat cake made with olive oil, pine nuts, walnuts and sultanas), pancakes and pasta.

Chestnuts are very much a seasonal food, with the chestnut harvest typically taking place during September, October and November, hence their availability over the Christmas period. Although chestnuts can be eaten raw, they are usually cooked before eating, either boiled or roasted. During the nineteenth century in the UK, roast chestnuts were a popular street food. In *London Labour and the London Poor*, Mayhew wrote of chestnut-selling as the oldest of the 'public traffics' and one which was widely practised, with every street fruit-seller also selling roast chestnuts. He observed:

> *The ordinary street apparatus for roasting chestnuts is simple. A round pan, with a few holes punched in it, costing 3d or 4d in a marine-store shop, has burning charcoal within it, and is surmounted by a second pan, or kind of lid, containing chestnuts, which are thus kept hot.*

Nowadays, chestnut stuffing is a popular accompaniment for roast turkey at Christmas. In his popular Victorian cookbook, *The Modern Cook* (1846), the Anglo-Italian chef Charles Elmé Francatelli, who had worked as Chief Cook and Maitre d'Hotel for Queen Victoria, gave a December menu which includes 'roast turkey, a la Chipolata', in which the bird is stuffed with chestnuts and served with sausages.

One classic use of chestnuts is to transform them into the delicacy known as *marrons glacés*. These are made in a process similar to that used in candying fruit. Peeled fresh chestnuts are steeped in an increasingly concentrated sugar syrup over a period of days so that the syrup saturates the nuts, with the sugar acting as a preservative as well as a sweetener. The resulting chestnuts have a delicate, frosted appearance from the sugar, as the name 'iced chestnuts' suggests. The time taken to produce *marrons glacés* is reflected in the high price they command. They are very much a luxurious treat, often given as a present during the festive season.

Marzipan

Marzipan, the name given to a paste made from ground almonds and sugar, has a long history of being served at feasts and celebrations. Sugar was for centuries a costly ingredient, used to create special treats, with marzipan accordingly very much a luxury. Nowadays, it is the cost of almonds rather than that of sugar which makes marzipan expensive.

One aspect of marzipan that has long been appreciated by confectioners and pastry chefs is its malleable aspect, which enables it to be shaped into decorative and symbolic shapes. During medieval times it was used to make the eye-catching models called 'solteties' (subtleties), which were brought in to mark the ends of courses during feasts. In Italy in 1470, the acclaimed Renaissance artist Leonardo da Vinci created marzipan sculptures for Ludovico Sforza, later the duke of Milan, observing ruefully in a codex that his creations had been entirely 'gobbled up'. The tradition of using marzipan to make colourful model fruits and animals is still found in many parts of Europe, such as in Portugal and on the Italian island of Sicily.

Marzipan continues to be very much prized in many countries. In Germany, the city of Lübeck has for centuries been noted for the quality of its marzipan. As it was the capital of the Hanseatic League (a medieval trade alliance), its confectioners had easy access to almonds and sugar and became famous for producing this costly delicacy. Today, the production of Lübeck marzipan is legally protected, with a high proportion of almonds being one of the stipulations in its production.

Toledo in Spain is another European city with a special association with marzipan, with its local manufacture also legally protected by a PGI. One Spanish food story makes the claim that marzipan was invented in Toledo. According to the legend, it was created by the nuns of the Convento de San Clemente when the city was besieged by the Arabs and they crushed almonds with sugar with a mace in order to help feed the hungry population. This was called *pan de maza* ('mace bread'). To this day, *mazapán de Toledo* continues to be made in the city in the form of *figuritas*, shapes such as stars, hearts or birds, which are then glazed and browned under a grill. One striking Christmas speciality from Toledo is the marzipan eel, a stylised, heraldic-looking eel made from marzipan and filled with candied egg yolks.

In northern Europe one finds the endearing tradition of giving marzipan pigs in a number of countries. In Denmark, as part of the Christmas Eve meal, a rice pudding called *risalamande* is served with a whole almond hidden in it. Whoever finds the almond is given a marzipan pig as a reward. In Germany, marzipan pigs are offered as gifts to bring good luck in the new year.

In the UK, the custom of creating marzipan subtleties fell out of fashion, but a taste for marzipan lingered on. It became an element of special-occasion cakes, such as the wedding cake and the Easter simnel cake. To this day, a layer of marzipan remains an expected component of a classic Christmas fruit cake, discreetly hidden under snow-white fondant or royal icing.

Christmas fruit cake

A celebratory Christmas cake – a rich fruit cake flavoured with spices, 'fed' with brandy or rum and topped with marzipan and snow-white icing – is now considered a classic British Christmas food. The Caribbean, too, has a tradition of Christmas fruit cake, including black cake, a dark fruit cake made from dried fruits soaked in rum and caramel syrup. The origins of eating fruit cake over the festive season in the UK can be traced back to the Twelfth cake, which was eaten to celebrate Twelfth Night, the end of the Twelve Days of Christmas. The Twelfth cake is a fruit cake, descended from the rich spiced and fruited breads enjoyed as a treat in medieval times. An ancestor is the 'plum' cake, with plum a term used to describe dried fruits.

Among traditional Twelfth Night customs is one thought to date back to the boisterous Roman festival of Saturnalia, held in December. During this period of merrymaking, roles were reversed, with slaves waited on by their masters. The idea of inverting roles became part of Christmas celebrations. Henry VII (r.1485–1509) had a 'Lord of Misrule' to preside over celebrations during the 12 days. Historically, the Twelfth cake contained a token hidden inside it, usually a dried bean. The person who found this token in their slice of cake became the Lord of Misrule or the bean king. One finds similar traditions in other parts of Europe, including France's *galette des rois* (layered with almond paste), which contains one little figurine or *fève;* and Spain's *roscón de reyes*, a fragrant sweet bread decorated with glacé fruits and nuts in which a figurine of a king or the baby Jesus is hidden – both of these are eaten to celebrate Epiphany.

In a charming piece of London theatre tradition, 6 January sees the cutting of the Baddeley cake at Drury Lane Theatre. The cake is named after Robert Baddeley (1733–1794), an actor and pastry chef who left a bequest in his will that Twelfth cake and punch be enjoyed by the company on Epiphany. It is a ceremony that was first carried out in 1795 and continues to this day.

As part of their Christmas celebrations while stuck in the ice on their expedition to Antarctica, Joseph Hooker and the HMS *Erebus* crew consumed a Twelfth cake on Twelfth Night which they all declared to be good, despite it being three years old.

Cranberry (*Vaccinium macrocarpon*) from *Gartenflora*, 1871.

The cranberry

The cranberry plant (*Vaccinium macrocarpon*) is native to the United States and Canada. A low-growing, evergreen shrub, it belongs to the same genus as the blueberry, the bilberry and the lingonberry and, like these, is valued for its edible fruit. Its natural habitat is wetlands such as bogs, fens or marshes. The Native Americans ate cranberries fresh and dried, adding them to pemmican (a pressed cake made from dried meat and fat), and they made tea from their leaves. It is thought that Native Americans introduced the European colonists to cranberries. Cranberries are noticeably tart, owing to the tannins they contain. They are high in vitamin C and were eaten to protect against scurvy. Today, the US is the world's leading producer of cranberries.

Cranberries ripen in the autumn and are harvested between September and November. While some cranberries are dry

harvested, the majority of the crop is wet harvested, a method developed in the 1960s which revolutionised cranberry farming. The bogs where the cranberries grow are flooded and the cranberries are dislodged from the plants by agitating the water. The red cranberries, which are naturally buoyant, float to the surface (creating a striking sight) and are then gathered together and collected.

As they are naturally tart in flavour, one popular way of preparing cranberries was to cook them into a sauce sweetened with sugar. Amelia Simmons's *American Cookery* (1796) – the first American cookbook written by an American and published in America – contains a recipe for cranberry sauce. In the US, cranberry sauce became a traditional accompaniment to roast turkey at the Thanksgiving meal (the family feast held in November to commemorate the first harvest of the Plymouth Colony in 1621). Delia Smith, the hugely influential, bestselling cookbook author and TV broadcaster, is credited with popularising cranberry sauce in the UK through her TV series and her bestselling Christmas cookbook of 1990. In the UK, cranberry sauce, either shop-bought or home-made, is now a popular complement to the Christmas roast turkey.

Citrus fruit

The cold, dark winter months of the northern hemisphere are the time of year when citrus fruit is at its best and most abundant. It has become an appealing tradition to place a satsuma, tangerine or clementine inside a Christmas stocking; the bright orange colour of the fruit having become associated with gold. Candied citrus peel – made from fruit including oranges (*Citrus × aurantium*), lemons

Clementine (*Citrus* x *aurantium*) from *Revue Horticole*, 1902.

(*Citrus* × *limon*) and citrons (*Citrus medica*) – is used as a flavouring in many of the traditional treats we enjoy eating at Christmas, from mince pies and Christmas cake to Italy's panettone and panforte.

Citrus is a genus of flowering trees and shrubs in the Rutaceae family. Native to a wide region of Asia and Oceania, citrus fruits are now cultivated in many countries around the world and are an important economic crop. Brazil, China and the United States are among the major citrus producing countries. An interesting aspect of the *Citrus* genus is its capacity to hybridise easily between species. Five wild ancestral species have been identified: the pomelo (*Citrus maxima*), the mandarin (*Citrus reticulata*) the citron (*Citrus medica*), the kumquat (*Citrus japonica*) and the makrut lime (*Citrus hystrix*). Natural crossing and mutations in the wild, together with selective breeding by citrus growers over the centuries, have resulted in a complex web of numerous citrus cultivars. The grapefruit for example, originated in the West Indies during the eighteenth century as a cross between the sweet orange (*Citrus* x *aurantium* Sweet Orange Group) and the pomelo.

Citrus fruits are valued for their juice, their fragrant peel and as a useful source of vitamin C. There are many culinary uses for citrus fruit. Marmalade, that very British breakfast conserve, is usually made from Seville oranges, a cultivar of sour orange, (*Citrus* x *aurantium* Sour Orange Group) but also can be made from other citrus fruits such as lemons or limes. Blood oranges, other cultivars in the Sweet Orange Group, have a distinctive red hue to their flesh and juice owing to the presence of anthocyanin pigments. Such is the concentration of flavour within a citrus fruit that a little goes a long way – think of the way that a slice of lemon perfumes a gin and tonic or a grating of lemon zest will lift a dish.

When citrus fruits were introduced into Europe, they were initially an exotic novelty, valued principally for their fragrance and used to scent rooms and linen. The citrus family grows naturally in tropical, subtropical or mild temperate climates.

In northern Europe by the sixteenth century, it had become fashionable among the aristocracy and wealthy merchants to have special orangeries – panelled with panes of expensive glass – built to display this prestigious, delicate new fruit. At Kew Gardens in 1761, the architect Sir William Chambers constructed the spacious Orangery in the classic style specifically so that citrus fruits could be cultivated inside it. In 1863, however, there was a change of use since the light levels inside were inadequate to sustain a healthy collection of citrus trees and the Orangery was repurposed as a Timber Museum. Nowadays, the elegant building houses a large cafe, busy with visitors to the Gardens. The citrus collection can today be found in the Temperate House.

The pomegranate

One of the seasonal fruits enjoyed during the winter months in Europe is the pomegranate, the fruit of the pomegranate tree (*Punica granatum*). Botanically speaking, the pomegranate is a berry, that is to say, a fleshy fruit – typically containing several seeds – that comes from a single flower with one ovary. Native to an area reaching from north-east Turkey to Afghanistan, the pomegranate tree has long been cultivated in India, the Middle East and the Mediterranean region. It's a striking fruit in appearance – a beautiful, pinkish-red globe that makes a fine addition to any fruit bowl. Breaking its hard shell open reveals the intricate clusters of closely packed, jewel-like, scarlet seeds inside. In fact, the name of the fruit comes from the Latin *pomum granatum*, which means apple of many seeds or grains. These glowing seeds make a lovely garnish for dishes such as dips and salads. Rich in polyphenols, the pomegranate enjoys a reputation as a healthy fruit. Pomegranate juice is a popular way of consuming it. Pomegranate molasses, made by cooking down the juice with sugar and lemon juice until it reduces to a thick, dark syrup, is a popular condiment in Middle Eastern cuisine. It adds a pleasant sour tang to dishes such as Iran's *fesenjoon* (a rich chicken and walnut dish) or Turkey's bulgur wheat salad *kisir*. In India, dried pomegranate seeds are turned into a spice called *anardana*, used to add a sweet–sour flavour to dishes.

The pomegranate is a fruit with much folklore attached to it in countries around the world. In Chinese culture, where red is an auspicious colour that symbolises fertility and vitality, the pomegranate is a symbol of fertility. In Buddhism it is one of the three blessed fruits, along with the peach and citrus varieties, and the Buddha is depicted holding these fruits. In Japan, the Buddhist deity Kishimojin, a goddess of easy birth and protector of children, is also often shown holding a pomegranate. In Greek mythology, the pomegranate plays an important part in the myth of Persephone, the beautiful daughter of Zeus and Demeter (the goddess of harvest). Persephone is abducted by Hades, god of the underworld, and kept underground in his realm. When Persephone's grieving mother threatens the world with famine, her father agrees that she could return to the surface, but only on condition that she hasn't eaten any food while in the underworld. However, as Persephone has been tempted to eat a few pomegranate seeds, she has to spend some months each year in the underworld. This period becomes winter, as Demeter mourns her daughter, while Persephone's return to the surface of the Earth is marked by the arrival of spring.

In Christianity, the pomegranate is associated with plenty. It is one of the fruits that Moses tells his followers will be found in the Promised Land – 'a land of wheat and barley, of vines and fig trees and pomegranates, a land of olive oil and honey' (Deuteronomy 8.8). It is considered a symbol of resurrection and

Pomegranate (*Punica granatum*) from *Pomologie Française*, 1846.

Grenadier à fruit doux.

22

De l'Imprimerie de Langlois — Bouquet Sc.

features in paintings of the Virgin Mary and Child. In Armenia, the pomegranate is held in high regard as a symbol of the country. At Armenian weddings, the bride breaks open a pomegranate by throwing it, with its scattered seeds representing her future children. In Greece, where the pomegranate has long been a symbol of abundance and fertility, it has become a custom to hang a pomegranate over the front door at Christmas and smash it on the doorstep as New Year's Day arrives, in order to bring good fortune to the household.

Sorrel (*Hibiscus sabdariffa*) from *Flore pittoresque et médicale des Antilles*, Michel Étienne Descourtilz, 1821.

Sorrel (hibiscus or roselle)

In the islands of the Caribbean, Christmas is celebrated with a striking, scarlet-hued drink called sorrel, named after the plant from which it is made. During the festive season, the drink is often enjoyed with rum, but it can also be served on its own, without any alcohol added. The sorrel plant in question is a type of hibiscus known as *Hibiscus sabdariffa*, not to be confused with the salad herb sorrel (*Rumex acetosa*). The fleshy red calyxes of the hibiscus flowers are used both fresh and dried to make a range of beverages, desserts such as sorbets and jam.

The drink sorrel is made by infusing the calyxes with water, spices and flavourings such as ginger, and sweetening it to taste. Sorrel has a noticeably tart, refreshing flavour, similar to cranberries. Part of sorrel's appeal is also an aesthetic one, as the flowers give drinks or desserts a beautiful dark-red colour.

Gingerbread

Among the traditional sweet treats enjoyed during the Christmas season are various spiced biscuits, reflecting the historic luxurious role of spices in the medieval kitchen. In Belgium and the Netherlands, *speculaas* biscuits flavoured with a mixture of spices are traditionally baked for St Nicholas Day in early December. Special, relief-carved wooden *speculaas* moulds are used to create decorative shapes such as birds, deer, windmills and the bearded figure of St Nicholas himself. In Germany, one finds *Adventgebäck* – an array of special biscuits for the season of Advent. Among these are traditional *lebkuchen,* honey biscuits, flavoured with a mixture of spices including ginger, which have their origins in monastic baking. The first *lebkuchen* maker was recorded in Nuremberg in the fourteenth century, and since then the city has become noted for them. The fact that the city was a key trading centre allowed its bakers access to the precious spices needed for them. Heart-shaped *lebkuchen* have become very popular at Christmastime, but one can also find them shaped as stars, Christmas trees, angels or bells.

Gingerbread – flavoured with aromatic ground ginger (see page 80) – is very much associated with Christmas. Gingerbread is made in numerous countries in an array of shapes and textures. In medieval times, it was made from honey mixed with breadcrumbs and spices. Powdered sandalwood (*Santalum album*) was sometimes added in order to give a reddish colour. Over the centuries the recipe for gingerbread changed in some countries, with treacle used rather than the historic honey, and flour instead of breadcrumbs. In France, however, the historic confection *pain d'épices,* eaten particularly at Christmastime, is classically still made from honey. In 1838, Queen Victoria was given gingerbread as a gift at Christmas by her mother, a German princess. Victoria's

German husband, Prince Albert, similarly embraced the tradition, dressing as St Nicholas and handing out gingerbread to their children.

Gingerbread's capacity to be shaped has played a part in its popularity. Gingerbread shapes known as fairings were popular treats sold at fairs in England. The fact that gingerbread can be shaped into gingerbread men has given it a special place in our affections. The spiced biscuit appears in children's stories such as the perennially popular *The Gingerbread Man*, while in Tchaikovsky's Christmas ballet *The Nutcracker*, the dashing nutcracker leads an army of gingerbread soldiers in a battle against the mice. Gingerbread shapes became popular decorations for Christmas trees. Another festive use for gingerbread is in creating the gingerbread house, a practice thought to have originated in Germany. The idea of a house made from gingerbread and decorated with icing captured the popular imagination with the fairy tale *Hansel and Gretel*, by the Brothers Grimm, in which two children are lured into a witch's magical gingerbread cottage. In Sweden, making and decorating a *pepparkakshus* (gingerbread house) has become an enjoyable Christmas family pastime. It is a tradition with a wide-ranging appeal. Since the 1960s, the White House in Washington DC has seen the creation of elaborate gingerbread houses by its pastry chefs for the enjoyment of the First Family and visitors at Christmastime.

Cacao and chocolate

Nowadays, chocolate is among the treats we enjoy during the festive period. Nets of golden chocolate coins are found in the toes of stockings, foil-wrapped chocolate baubles decorate Christmas trees, and rich chocolate truffles are served as an after-dinner indulgence. Historically, however, chocolate was for a long time consumed primarily as a beverage, rather than eaten in solid form. Chocolate is made from the seeds (known as beans) found inside the fruit of the cacao tree (*Theobroma cacao*). The tree is native to South America but is now cultivated in a number of tropical countries including Ghana, Nigeria and Indonesia. Cacao trees thrive in areas with high rainfall and humidity and do best in shade. Whereas most plants flower by growing new stalks, cacao is unusual as it flowers on its trunk and flowers best on wood that is over three years old. This botanical trait is called cauliflory. In the wild, these flowers, which usually only last a day, are pollinated by tiny insects. At Kew Gardens, where there are cacao trees growing in both the Palm House and the Princess of Wales Conservatory, the flowers are hand pollinated by horticulturists. The cacao fruit, called a pod, is shaped like a rugby ball and ripens into a variety of colours, including green, yellow, orange, red and purple. The pods are filled with cacao beans and a sweet white paste. These beans were valued by early Mesoamerican civilisations, including the Olmecs and the Mayans, who ground the beans to make beverages as well as using them as currency.

Cacao was also highly valued in the Aztec empire. Cacao beans were sent as tributes to the Aztec rulers. In Aztec society, *cacáhuatl* (chocolate water) was a drink enjoyed by the elite, including the emperor. It was a ceremonial drink, served at banquets and offered to the gods. Cacao is naturally astringent and the drink made from it was flavoured in many ways, with additions including chilli, honey and vanilla. In the sixteenth-century *Florentine Codex*, a detailed record of Aztec society, there is a fascinating description given by the Spanish friar Bernardino de Sahagún of how an Aztec seller made the drink:

> *She grinds cacao, she crushes, breaks, pulverises them. She chooses, selects, separates them. She drenches, soaks, steeps them. She adds water sparingly, conservatively; aerates it, filters it, strains it, pours it back and fourth, aerates it; she makes it form a head, makes foam; she removes the head, makes it thicken, makes it dry, pours water in it, stirs water into it.*

Following the conquest of the Aztec empire by the Spanish empire (1519–1521), chocolate in its beverage form was introduced to Europe. It is often said that the word 'chocolate' derives from the Nahuatl word *cacáhuatl*, but written records do not conclusively prove this. Reactions to the novel, strange, dark brew varied, with Girolamo Benzoni in his *History of the New World* (1565) observing, 'It seemed more a drink for pigs, than a drink for humanity.' Others, however, acquired a taste for this drink. Chocolate was deemed to have both beneficial health effects and aphrodisiac properties and was appreciated for these. The fact that it was a drink which did not cause inebriation was also valued. Just as in Aztec society this exotic drink had been enjoyed by the elite, so it was in Europe, where the wealthy and powerful considered it very much an expensive luxury. The Spanish court in the seventeenth century was noted for its hot chocolate, and the fashion spread among other European royalty and aristocracy. In Florence, Cosimo III de' Medici's physician, Francesco Redi, invented a renowned jasmine chocolate drink for his patron, infusing the cacao with fragrant fresh jasmine flowers (*Jasminum officinale*). In around 1690, as part of William III and Queen Mary II's rebuilding of Hampton Court Palace, the architect Christopher Wren built special Chocolate Kitchens there. It was here that royally appointed hot chocolate makers would prepare chocolate beverages for the king and queen, usually at breakfast time. In contrast, in 1693 at the Palace of Versailles, King Louis XIV supressed the serving of chocolate for his guests on grounds of economy.

Cacao (*Theobroma cacao*), 1820.

Chocolate continued to find a receptive audience. In England, coffee houses served chocolate in the form of a drink, as well as coffee. The diarist Samuel Pepys was always interested in trying what was new and fashionable. Writing in 1664, he noted, 'About noon out with Commissioner Pett, and he and I to a Coffee-house, to drink jocolatte, very good.'

The chocolate that we enjoy today in solid form rather than as a drink came about as a result of the Industrial Revolution in Europe. Cacao is naturally rich in a fat called cacao butter. A key invention in the history of chocolate is Dutchman Coenraad Johannes van Houten's development of a screw press, patented in 1828, which removed two-thirds of the cacao butter from the cacao solids, leaving behind a 'cake' which could be ground to a powder called cocoa. So that the cocoa would mix well with water, Van Houten alkalised it, a process that became known as 'Dutching'. Following on from this discovery, in 1847 the Quaker firm of J.S. Fry & Sons in Bristol enterprisingly mixed together cocoa powder, sugar and some of the extracted cacao butter to produce a paste which could be moulded into bars. These were grandly named 'chocolat deliceux à manger' (chocolate that's delicious to eat), and this was a key moment in the history of solid chocolate. Another important development came about in 1879, when Swiss chocolate manufacturer Rudolphe Lindt invented a machine called a conche and a process known as 'conching'. This was a way of mechanically moving the chocolate mass, reducing the particle size and creating smooth chocolate. The same year also saw the creation of milk chocolate, with powdered milk (another recent invention) mixed into the mixture.

The confectionery industry embraced the new solid chocolate with enthusiasm and inventiveness. The fact that tempered chocolate can be moulded saw the creation of treats such as liqueur chocolates, truffles and chocolate-coated nuts and dried fruits. The Christmas season sees specialist chocolatiers offering tempting ranges of seasonally flavoured truffles and appealing chocolate shapes from penguins to Christmas trees.

[following page]
Cacao (*Theobroma cacao*) from *Köhler's Medizinal-Pflanzen*, FranzEugen Köhler, 1887.

RECIPES

Stilton and poppy seed coins

Makes approx. 40 biscuits

These moreish, coin-shaped cheesy biscuits are a very good way of using up some of your Christmas Stilton. Serve with drinks as an appetiser and watch them disappear!

75g plain flour, plus extra for flouring the surface

1 tsp baking powder

50g butter, diced, plus extra for greasing

75g Stilton, finely crumbled

25g Cheddar, finely grated

15g poppy seeds

1 egg yolk

Preheat the oven to 180°C/160°C fan. Grease and line two baking sheets with baking paper.

In a mixing bowl, place the flour and baking powder and mix together. Using your fingertips, rub in the butter until absorbed.

Add the Stilton, Cheddar and poppy seeds, stirring to mix in well. Add the egg yolk and mix in to form a soft, sticky dough.

Divide the dough into two even-sized portions for easier handling. On a lightly floured surface, roll each portion into a cylinder shape approximately 3cm in diameter. Cut each cylinder into slices 0.5cm thick. Place the slices, spaced apart, on the prepared baking sheets.

Bake for 10–15 minutes until pale golden, then remove from the oven and allow to cool and set. Store in an airtight container.

Spiced sweet potato soup

Serves 4

Aromatic ginger and fragrant cinnamon and nutmeg give a delightful lift to sweet potatoes in this comforting soup. It's perfect for a light lunch or as a warming Thermos flask soup to enjoy on a long winter's-day walk.

- 600g sweet potatoes
- splash of lemon juice or vinegar
- 1 tbsp oil
- 1 small onion, finely chopped
- 5mm root ginger, peeled, finely chopped
- ½ tsp ground ginger
- ½ tsp ground cinnamon
- 800ml vegetable (or chicken) stock, or water
- salt and black pepper
- freshly grated nutmeg
- natural yogurt or soured cream, to garnish
- chopped coriander leaves or parsley, to garnish
- bread, to serve

Peel the sweet potatoes, chop into small chunks and pop these into a pan or bowl of water acidulated with a little lemon juice or vinegar to prevent discoloration. Set aside.

Heat the oil in a large saucepan. Add the onion and ginger and cook over a low heat, stirring often, for 5 minutes. Sprinkle in the ground ginger and cinnamon, and mix. Drain the sweet potatoes and add to the pan, stirring to coat the chunks with the spices.

Add the stock or plain water, and bring to the boil. Reduce the heat and simmer, uncovered, for 25 minutes, until the sweet potatoes are tender.

Blend the soup until smooth, then return it to the pan. Season with salt and black pepper plus a generous grating of nutmeg.

Gently heat through until piping hot. Serve garnished with natural yogurt or soured cream plus chopped coriander leaves or parsley, with bread on the side.

Just before serving, stir in the pomegranate seeds well.

Chestnut soup

Serves 4–6

Chestnuts add their subtle yet distinctive flavour to this rich, textured soup. Serve with bread for a first course or a light lunch.

- 1 tbsp olive oil
- 15g butter
- 1 onion, finely chopped
- 1 leek, halved lengthways and finely sliced
- 2 celery stalks, finely sliced
- 1 bay leaf
- 400g cooked, peeled chestnuts, roughly chopped
- splash of Madeira or Amontillado (optional)
- 800ml chicken or vegetable stock
- 1 tbsp finely chopped parsley
- salt and freshly ground black pepper
- freshly grated nutmeg
- chopped parsley, to garnish

In a large saucepan, heat the olive oil and butter over a medium heat. Add the onion, leek, celery and bay leaf and fry gently over a low heat for 10 minutes, stirring often, until the onion and leek are softened.

Add the chestnuts, mixing in. If you are using Madeira or Amontillado, add it now and cook over a medium heat for 2–3 minutes to partially reduce the alcohol.

Add the stock and parsley, bring to the boil, reduce the heat, cover partly and simmer gently for 20 minutes. Season with salt and black pepper and add some nutmeg.

Discard the bay leaf.

Remove half of the mixture from the pan and blend it until smooth. Return it to the pan and mix it with the remaining textured half.

Simmer gently to heat through. Serve garnished with parsley.

***Chestnutting* by Winslow Homer from *Every Saturday: An Illustrated Journal of Choice Reading*, 1870.**

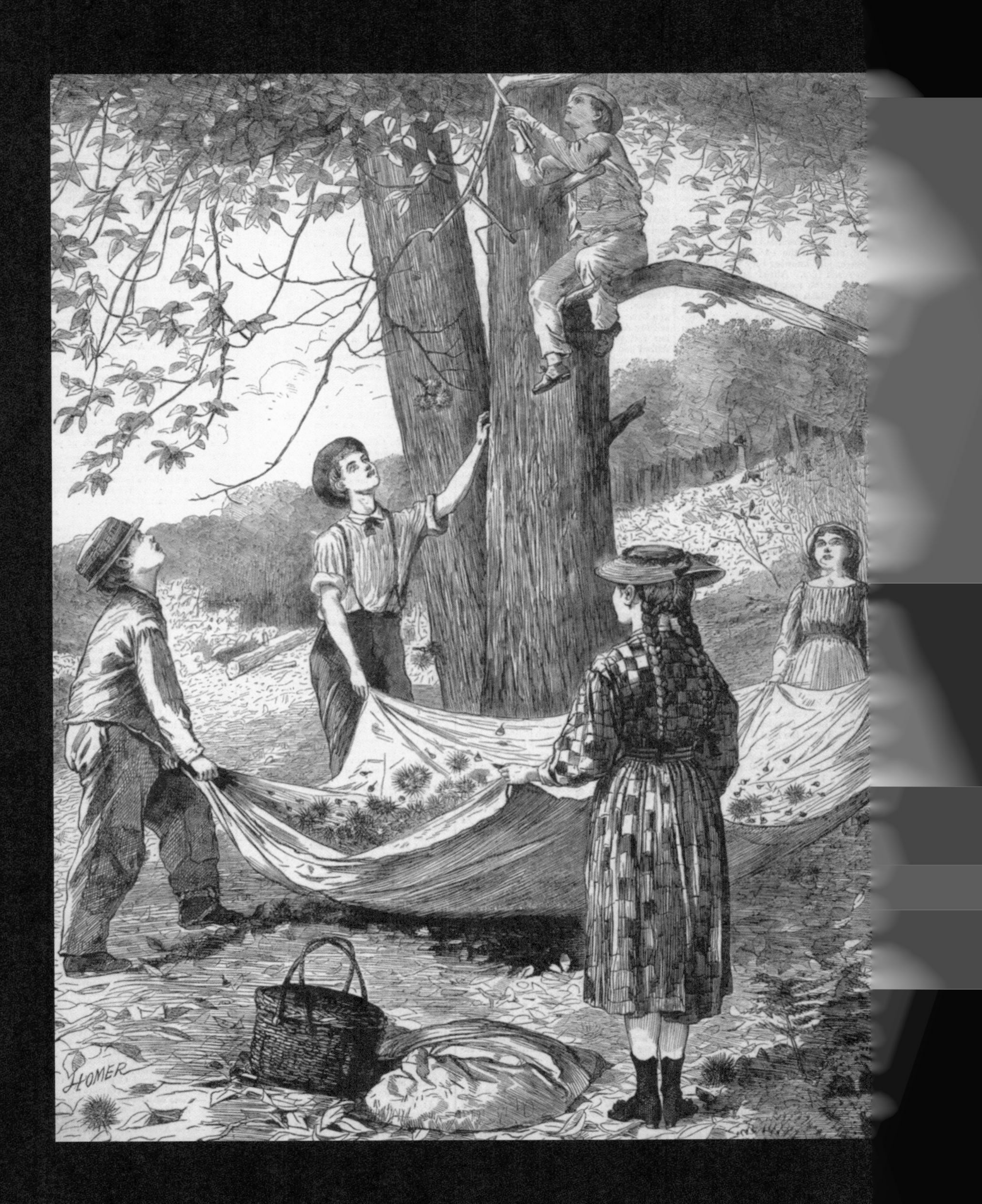
HOMER

Crunchy pickled vegetables

Makes 1 x 500g jar

These pickled vegetables are simple to prepare at home and deliver good results, being crunchy, pleasantly tangy and fragrant from the dill and spices. Serve as an enlivening accompaniment to a festive cheeseboard, cold roast meats or a charcuterie platter.

- 100g carrots, cut into short fine matchstick strips
- 150g cauliflower florets, finely sliced
- 50g shallots, peeled, halved and finely sliced
- 1 tsp salt
- 125ml cider vinegar or white wine vinegar
- 125ml water
- 30g light brown sugar
- ¼ tsp celery seeds
- ½ tsp caraway seeds
- 2 tbsp finely chopped dill

In a large bowl, mix together the carrot, cauliflower, shallots and salt. Set aside for one hour.

Just before the hour is up, place the vinegar, water, sugar, celery seeds and caraway seeds in a saucepan and bring to the boil. Pour the hot vinegar mixture over the vegetables. Stir in the dill.

Transfer the pickles to a sterilised jar, allow to cool, cover, then refrigerate. The pickles will keep in the fridge for a month.

Beetroot potato pancakes

Makes approx. 16 pancakes

400g raw beetroot, peeled
200g potatoes, peeled
1 small red onion, finely chopped
115g plain flour
2 eggs, beaten
salt and freshly ground black pepper
oil, for frying
soured cream, to serve
apple sauce, to serve

Beetroot gives a striking colour and earthy sweetness to these potato pancakes. Serve them as part of a brunch or lunch dish, perhaps with slices of Christmas ham.

Grate the beetroot and potatoes. Wrap the grated vegetables in a clean tea towel or similar cloth and squeeze out the excess moisture.

Place the grated vegetables in a large bowl. Add the onion, flour and eggs and mix together well. Season with salt and freshly ground pepper.

Pour enough oil into a large frying pan to form a shallow layer. Heat it through over a medium heat. Add four separate tablespoons of the beetroot mixture, spacing them well apart. With a spatula, flatten down each mound of mixture to form a roughly circular pancake, about 5mm thick. Fry for 2–3 minutes until nicely golden brown underneath, then carefully flip over each pancake and fry for 2–3 minutes more on the other side. Remove the fried pancakes from the pan, drain on kitchen paper and keep them warm.

Repeat the process with the remaining mixture until it has all been used up, adding more oil to the pan if needed. Serve at once with soured cream and apple sauce.

Pomegranate coleslaw

Serves 4–6

A delightfully crunchy coleslaw is a very good addition to the Christmas table. It makes an excellent accompaniment to cold turkey, roast beef or ham. In this seasonal version, garnet-like pomegranate seeds add colour, sweetness and an appealing texture to this perennially popular cabbage salad.

- ½ white cabbage, finely sliced
- salt and freshly ground black pepper
- juice and grated zest of ½ lemon
- 3 tbsp mayonnaise
- 2 tbsp natural yogurt or crème fraiche
- 1–2 tsp wholegrain mustard
- 1 tsp olive oil
- 1 spring onion, finely chopped
- 1 celery stalk, finely sliced
- 25g walnut pieces, chopped small
- 6 tbsp finely chopped dill
- seeds of 2 small pomegranates or 1 large one, any bitter white pith carefully picked off and discarded

In a large bowl, place the sliced cabbage with salt to taste plus the lemon juice, and mix together well.

To make the dressing, in a separate bowl or jug mix together the mayonnaise, yogurt or crème fraiche, lemon zest, wholegrain mustard and olive oil. Season with freshly ground pepper.

To the cabbage, add the spring onion, celery and chopped walnuts and mix well. Pour on the dressing, add the dill and mix together thoroughly, making sure the salad is coated well with the dressing.

Just before serving, stir in the pomegranate seeds.

Truffled porcini stuffing

Serves 4–6

This vegetarian stuffing is a delicious celebration of fungi, combining the earthy flavours of dried porcini and fresh mushrooms with the distinctive taste of truffles.

- 10g dried porcini mushrooms
- 2 tbsp olive oil
- 1 onion, finely chopped
- 10g butter, plus extra for greasing
- 200g mushrooms, finely chopped
- splash of dry white wine
- salt and freshly ground black pepper
- 75g fresh breadcrumbs
- 1 large egg
- 1 tbsp finely chopped parsley
- 1 tsp grated lemon zest
- 1–2 tsp truffle oil

Grease a shallow baking dish and set aside.

Place the dried porcini mushrooms in a heatproof bowl, cover with boiling water and set aside to soak for 15–20 minutes. Strain and finely chop.

Heat 1 tbsp olive oil in a large, heavy frying pan. Add the onion and fry over a low heat for 10–15 minutes until softened.

Add the butter to the pan. Once the butter has melted, add the chopped fresh mushrooms and porcini mushrooms. Fry for 5 minutes, stirring often. Add the wine and fry for a further 5 minutes, stirring often. Season well with salt and freshly ground pepper. Transfer the mixture to a large bowl.

Add the breadcrumbs to the mushroom mixture and break in the egg. Mix together well. Stir in the parsley, lemon zest and truffle oil.

Transfer the stuffing to the greased baking dish. Drizzle with the remaining olive oil. Bake at 180°C/160°C fan for 25–30 minutes until golden brown. Serve warm from the oven.

Christmas plum pudding and other desserts from *Mrs Beeton's Cookery Book*, Isabella Mary Beeton, 1865.

nd Grapes.
Ice Pudding.
Lemon Jelly.
Cherries.
Melon and Green figs.
Candied Oranges.
Plums.
lding.
Ribbon Jelly.
Ices.
Meringues.
Sponge Cake.
Greengages.
Open Tart.
Plum Cake.
Compote of Apples.
Gâteau.

Cherry walnut Christmas mincemeat

Makes approx. 1.2kg

Making your own Christmas mincemeat is simple, and the results are satisfying. Try this version, enriched with nicely tangy dried cherries and chopped walnuts. It makes the most delicious mince pies!

250g Bramley apples

115g suet or vegetarian suet

125g pitted dried cherries

125g raisins

75g sultanas

75g currants

50g candied peel, finely chopped

175g soft dark brown sugar

grated zest and juice of 1 orange

grated zest and juice of 1 lemon

50g walnut halves, chopped

2 tsp mixed ground spice

½ tsp ground allspice

½ tsp ground cinnamon

a good grating of nutmeg

3 tbsp brandy or Calvados

Quarter and core the apples, then chop finely.

In a large ovenproof bowl, place the chopped apple, suet, dried fruits, candied peel, soft dark brown sugar, orange and lemon zest and juice, and walnuts. Mix together well. Add all the spices and mix in. Cover with foil and set aside overnight.

Preheat the oven to 120°C/100°C fan. Bake the mincemeat mixture in its bowl for 3 hours.

Remove from the oven and stir the mixture now and then as it cools, so as to mix the melted suet with the dried fruit and peel. Once the mincemeat is thoroughly cool, stir in the brandy or Calvados.

Transfer into sterilised jars, cover and store in a cool, dry cupboard, where it will keep for many months. Once opened, keep it in the fridge.

Cherry (*Prunus avium*) from *La Belgique Horticole*, 1853.

Panettone pudding

Serves 6

This luxurious dessert – a riff on that much-loved classic, bread and butter pudding – makes very good use of panettone and is perfect for Christmas!

- 300g panettone
- 125ml double cream
- 125ml whole milk
- finely grated zest of 1 orange
- 1 tsp vanilla extract
- 3 eggs
- 40g caster sugar
- 100g butter, melted
- 2 tbsp demerara sugar

Preheat the oven to 180°C/160°C fan.

Cut the panettone into slices and trim off the crusts.

In a saucepan, bring the cream, milk and orange zest to boiling point. Remove from the heat and stir in the vanilla extract.

In a large bowl, whisk together the eggs and caster sugar until well mixed. Pour on the hot cream and milk, whisk together well and then pass the mixture through a fine sieve.

Brush each panettone slice generously with melted butter. In an ovenproof serving dish, arrange the slices, overlapping them.

Pour over the egg and cream mixture as well as any remaining melted butter. Set aside to soak for 20 minutes. Sprinkle evenly with demerara sugar.

Bake in the oven for 30–40 minutes until golden brown. Serve warm from the oven.

Syllabub ginger trifle

Serves 10

Christmas wouldn't be Christmas without a trifle! This version combines two historic British desserts – the syllabub and the trifle – to splendid effect!

- 50 ml brandy
- 75g caster sugar
- 1 lemon, juice and grated zest
- 250g ginger loaf cake, sliced 1cm slices
- 3 tbsp sherry
- 3 pears, peeled, quartered, poached, sliced into 1cm thick slices
- 500g thick custard
- 600ml double cream
- 120ml sweet white wine
- 2–3 tbsp toasted flaked almonds

To prepare the syllabub topping, gently heat the brandy and sugar together, stirring until the sugar dissolves. Allow to cool and once cooled stir in the lemon juice and zest and set aside.

Make the trifle by layering the ginger cake slices in a large serving bowl. Pour the sherry over the cake. Top with the poached pears, then pour over the custard, so that it forms a layer. If making in advance, cover and chill at this stage.

Before serving, finish making the syllabub. Whisk the double cream until it just begins to hold its shape. Add the brandy and lemon mixture a little at a time, whisking in between. Add in the sweet white wine, again just a little at a time, whisking in between. Spread the syllabub over the custard in an even layer. Sprinkle evenly with the toasted flaked almonds and serve.

Sorrel rum jelly

Serves 6

Inspired by the sorrel rum punch enjoyed in the Caribbean over the Christmas period (see page 102), this tipsy, appealingly wobbly jelly is for the grown-ups. Garnet in colour and studded with orange tangerine or clementine segments, it makes a beautiful, elegant and refreshing dessert.

- 15g dried sorrel (hibiscus flowers)
- 1 clove
- 1 cinnamon stick
- 2 broad strips of orange peel
- 600ml boiling water
- 80g sugar
- 5 gelatin leaves
- 50ml rum
- 2 tangerines or clementines, peeled and separated into segments

First make the sorrel liquid: in a heatproof container, place the dried sorrel, clove, cinnamon stick and orange peel and pour over the boiling water. Add the sugar and stir until dissolved. Set aside to infuse and cool for 1–2 hours, then strain through a sieve.

Soak the gelatin leaves in a little cold water for 4–5 minutes to soften. Pour the sorrel liquid into a pan and heat it gently. Gently squeeze the soaked gelatin leaves to remove excess moisture, then add the softened gelatin to the simmering sorrel liquid and stir until it has dissolved. Stir in the rum.

Pour the sorrel liquid into a serving bowl. Add the tangerine or clementine segments, which will float to the surface. Set aside to cool. Cover, and chill in the fridge to set – for around six hours, or overnight.

Cranberry marzipan cookies

Makes approx. 50 cookies

Homemade cookies are always a treat. These almond-flavoured ones – studded with pieces of marzipan, almonds and nicely chewy cranberries – go very well indeed with a cup of coffee.

- 115g butter, softened
- 100g dark brown sugar
- 100g caster sugar
- 1 medium egg
- 2 tsp almond extract
- 200g plain flour
- 1 tsp bicarbonate of soda
- finely grated zest of 1 lemon
- 50g marzipan, finely diced
- 50g toasted almonds, chopped
- 75g dried cranberries

Preheat the oven to 180°C/160°C fan. Line some baking sheets with baking paper.

In a large mixing bowl, cream together the butter and the two types of sugar until well mixed.

Beat together the egg and the almond extract, and mix into the creamed butter and sugar.

Sift the flour and bicarbonate of soda and fold into the cookie dough mixture, until thoroughly incorporated. Mix in the lemon zest, diced marzipan, chopped almonds and cranberries.

Using your hands, shape pieces of the cookie dough into walnut-sized balls and place them, spaced well apart, on the baking sheets. Bake in the oven for 10–15 minutes until golden brown.

Remove from the oven, allow to cool and firm up slightly on the baking sheets, then cool on wire racks. Store in an airtight container.

Chocolate orange truffles

Makes approx. 36 truffles

Chocolate truffles are great fun to make and you can flavour them as you wish. These ones are delicately flavoured with fragrant citrus, spices and orange blossom water. Serve them as an after-dinner treat.

150g dark chocolate (70% cocoa solids), finely chopped

150ml double cream

25g unsalted butter

grated zest of 1 orange

1 tbsp orange juice

seeds of 3 cardamom pods, finely ground

½ tsp ground cinnamon

½ tsp orange blossom water

cocoa powder, sifted, for coating

Place the chopped chocolate in a heatproof bowl.

In a small saucepan, gently heat together the cream and butter until the butter melts and the mixture reaches simmering point. Be careful not to let the mixture come to the boil, as if it's too hot the truffle mixture will seize.

Pour the hot cream mixture over the chocolate and stir until the chocolate has melted. Stir in the orange zest and orange juice, the ground cardamom and cinnamon, and the orange blossom water.

Set aside to cool. Cover and chill for at least 6 hours or overnight.

To make the truffles, scoop out heaped teaspoons of the chocolate mixture and shape into small rounded balls. Roll the balls in sifted cocoa powder. Cover and chill before serving.

Fry's advertising leaflet, J. S. Fry & Sons, c. 1890s.

FRY'S Celebrated
Cocoa and
Chocolates.

Recommended reading

Simon Barnes. *The Green Planet: The Secret Life of Plants*. Witness Books, London (2022).

Bernd Brunner, trans. Benjamin A. Smith. *Inventing the Christmas Tree*. Yale University Press, New Haven and London (2012).

Alan Davidson. *The Oxford Companion to Food*. Oxford University Press, Oxford (1999).

Charles Dickens. *A Christmas Carol*. Chapman and Hall, London (1843).

Judith Flanders. *Christmas: A Biography*. Picador, London (2017).

Annie Gray. *At Christmas We Feast: Festive Food through the Ages*. Profile Books, London (2021).

Dorothy Hartley. *Food in England: A Complete Guide to the Food that Makes Us Who We Are*. Macdonald, London (1954).

Ronald Hutton. *The Stations of the Sun: A History of the Ritual Year in Britain*. Oxford University Press, Oxford and New York (1996).

Timothy Larsen, ed. *The Oxford Handbook of Christmas*. Oxford University Press, Oxford (2020).

Michael Leach and Meriel Lland. *A Natural History of Christmas: A Miscellany*. Quandary Press, UK (2013).

Laura Mason and Catherine Brown. *Traditional Foods of Britain: An Inventory*. Prospect Books, Totnes (1999).

Harold McGee. *On Food and Cooking: The Science and Lore of the Kitchen*. 2nd edn, Scribner, New York (2004).

Mark Nesbitt and Lydia White, eds. *Festive Flora*. Royal Botanic Gardens, Kew, Richmond (2021).

Jeanne L.D. Osnas and Katherine Angela Preston. 'The Botanist in the Kitchen', blog. botanistinthekitchen.blog

Andy Thomas. *Christmas: A Short History from Solstice to Santa*. Ivy Press, London (2019).

Herbert H. Wernecke. *Christmas Customs around the World*. Westminster Press, Philadelphia (1959).

Father Christmas and His Little Friends, 1880.

Image credits

Bridgeman Images: 22 (Look and Learn / Illustrated Papers Collection / Bridgeman Images)

Mary Evans Picture Library: 34 (below)

Mary Evans Picture Library / Peter & Dawn Cope Collection: 34 (above)

Rijksmuseum: 4, 6 (Gift of A. Allebé, Amsterdam); 8 (Gift of. Perrée, Eindhoven); 18, 67, 70 (Gift of F.G. Waller, Amsterdam)

RTRO / Alamy Stock Photo: 12–13

Shawshots / Alamy Stock Photo: 120–1

Smithsonian: 21 (Harry T. Peters "America on Stone" Lithography Collection); 42 (Gift of John Goldsmith Phillips, Jr.); 64–5 (Smithsonian American Art Museum, Gift of Laura Dreyfus Barney and Natalie Clifford Barney in memory of their mother, Alice Pike Barney); 115 (Smithsonian American Art Museum, The Ray Austrian Collection, gift of Beatrice L. Austrian, Caryl A. Austrian and James A. Austrian)

The Met: 30, 46 (Gift of Mrs. Richard Riddell, 1981); 53 (The Elisha Whittelsey Collection, The Elisha Whittelsey Fund, 1996); 76 (The Jefferson R. Burdick Collection, Gift of Jefferson R. Burdick); 131 (Gift of Mrs. Richard Riddell, 1981)

Wellcome Collection: 47; 50–1; 56; 75; 129

York Museums and Gallery Trust: 44

Publisher's acknowledgements

Kew Publishing would like to thank the following for their help and contribution: Bill Baker, Sara Barrios, Owen Blake, Aisyah Faruk, Tony Hall, Thomas Heller, Rory Hutton, David Mabberley, Elizabeth Mansfield, Kevin Martin, Mark Nesbitt, Richard Wilford.

Conversion tables

Abbreviations used in this book

g	gram
kg	kilogram
ml	millilitre
l	litre
mm	millimetre
cm	centimetre
tsp	teaspoon
tbsp	tablespoon

Weights

5g	¼ oz
15g	½ oz
20g	¾ oz
25g	1 oz
50g	2 oz
75g	3 oz
100g	4 oz
150g	5 oz
175g	6 oz
200g	7 oz
250g	8 oz
275g	9 oz
300g	10 oz
325g	11 oz
350g	12 oz
375g	13 oz
400g	14 oz
500g	1 lb
1kg	2 lb

US weight equivalents

25g	(1oz)	⅛ cup
50g	(2oz)	¼ cup
100g	(4oz)	½ cup
175g	(6oz)	¾ cup
250g	(8oz)	1 cup
500g	(1lb)	2 cups

Liquids / volumes

5ml	¼ fl oz*
15ml	½ fl oz
25ml	1 fl oz
45ml	1½ fl oz
50ml	2 fl oz
75ml	3 fl oz
100ml	3½ fl oz
125ml	4 fl oz
150ml	¼ pt
175ml	6 fl oz
200ml	7 fl oz
250ml	8 fl oz
275ml	9 fl oz
300ml	½ pt
350ml	12 fl oz
375ml	13 fl oz
400ml	14 fl oz
450ml	¾ pt
500ml	17 fl oz
600ml	1 pt
750ml	1¼ pt
900ml	1½ pt
1 litre	1¾ pt

*1 teaspoon

Measurements

5mm	¼ inch
1cm	½ inch
1.5cm	¾ inch
2.5cm	1 inch
5cm	2 inches
10cm	4 inches
15cm	6 inches
20cm	8 inches
25cm	10 inches
30cm	12 inches

Oven temperatures

140°C	275°F	gas mark 1
150°C	300°F	gas mark 2
170°C	325°F	gas mark 3
180°C	350°F	gas mark 4
190°C	375°F	gas mark 5
200°C	400°F	gas mark 6
220°C	425°F	gas mark 7
230°C	450°F	gas mark 8
240°C	475°F	gas mark 9

Index

First published in 2023 by
Royal Botanic Gardens, Kew,
Richmond, Surrey, TW9 3AB, UK
www.kew.org

ISBN 978 1 84246 793 0

Distributed on behalf of the Royal Botanic Gardens, Kew in North America by the University of Chicago Press, 1427 East 60th St, Chicago, IL 60637, USA.

British Library Cataloguing in Publication Data
A catalogue record for this book is available from the British Library

Design and page layout: Ocky Murray
Cover illustrations and internal motifs: Rory Hutton
Project management: Lydia White
Copy-editing: Marija Duric Speare
Proofreading: Jo Mortimer

Printed and bound in Great Britain by Gomer Press

For information or to purchase all Kew titles please visit shop.kew.org/kewbooksonline or email publishing@kew.org

Kew's mission is to understand and protect plants and fungi, for the wellbeing of people and the future of all life on Earth.

Kew receives approximately one third of its funding from Government through the Department for Environment, Food and Rural Affairs (Defra). All other funding needed to support Kew's vital work comes from members, foundations, donors and commercial activities, including book sales.